Environmental Management Systems: A Tool to Help Water Utilities Manage More Effectively

Awwa
Research
Foundation

About the Awwa Research Foundation

The Awwa Research Foundation (AwwaRF) is a member-supported, international, nonprofit organization that sponsors research to enable water utilities, public health agencies, and other professionals to provide safe and affordable drinking water to consumers.

The Foundation's mission is to advance the science of water to improve the quality of life. To achieve this mission, the Foundation sponsors studies on all aspects of drinking water, including supply and resources, treatment, monitoring and analysis, distribution, management, and health effects. Funding for research is provided primarily by subscription payments from approximately 1,000 utilities, consulting firms, and manufacturers in North America and abroad. Additional funding comes from collaborative partnerships with other national and international organizations, allowing for resources to be leveraged, expertise to be shared, and broad-based knowledge to be developed and disseminated. Government funding serves as a third source of research dollars.

From its headquarters in Denver, Colorado, the Foundation's staff directs and supports the efforts of more than 800 volunteers who serve on the board of trustees and various committees. These volunteers represent many facets of the water industry, and contribute their expertise to select and monitor research studies that benefit the entire drinking water community.

The results of research are disseminated through a number of channels, including reports, the Web site, conferences, and periodicals.

For subscribers, the Foundation serves as a cooperative program in which water suppliers unite to pool their resources. By applying Foundation research findings, these water suppliers can save substantial costs and stay on the leading edge of drinking water science and technology. Since its inception, AwwaRF has supplied the water community with more than $300 million in applied research.

More information about the Foundation and how to become a subscriber is available on the Web at **www.awwarf.org.**

Environmental Management Systems: A Tool to Help Water Utilities Manage More Effectively

Prepared by:
CH2M HILL
13921 Park Center Road, Suite 600
Herndon, VA 20171

Jointly sponsored by:
Awwa Research Foundation
6666 West Quincy Avenue, Denver, CO 80235-3098
and
U.S. Environmental Protection Agency
Washington D.C.

Published by:

Awwa Research Foundation

American Water Works Association

IWA Publishing

DISCLAIMER

This study was jointly funded by the Awwa Research Foundation (AwwaRF) and the U.S. Environmental Protection Agency (USEPA) under Cooperative Agreement No. R-829679-01. AwwaRF and USEPA assume no responsibility for the content of the research study reported in this publication or for the opinions or statements of fact expressed in the report. The mention of trade names for commercial products does not represent or imply the approval or endorsement of AwwaRF or USEPA. This report is presented solely for informational purposes.

CONTENTS

TABLES

FIGURES

FOREWORD

The Awwa Research Foundation is a nonprofit corporation that is dedicated to the implementation of a research effort to help utilities respond to regulatory requirements and traditional high-priority concerns of the industry. The research agenda is developed through a process of consultation with subscribers and drinking water professionals. Under the umbrella of a Strategic Research Plan, the Research Advisory Council prioritizes the suggested projects based upon current and future needs, applicability, and past work; the recommendations are forwarded to the Board of Trustees for final selection. The foundation also sponsors research projects through the unsolicited proposal process; the Collaborative Research, Research Applications, and Tailored Collaboration programs; and various joint research efforts with organizations such as the U.S. Environmental Protection Agency, the U.S. Bureau of Reclamation, and the Association of California Water Agencies.

This publication is a result of one of these sponsored studies, and it is hoped that its findings will be applied in communities throughout the world. The following report serves not only as a means of communicating the results of the water industry's centralized research program but also as a tool to enlist the further support of the nonmember utilities and individuals.

Projects are managed closely from their inception to the final report by the foundation's staff and large cadre of volunteers who willingly contribute their time and expertise. The foundation serves a planning and management function and awards contracts to other institutions such as water utilities, universities, and engineering firms. The funding for this research effort comes primarily from the subscription program, through which water utilities subscribe to the research program and make an annual payment proportionate to the volume of water they deliver and consultants and manufacturers subscribe based on their annual billings. The program offers a cost-effective and fair method for funding research in the public interest.

A broad spectrum of water supply issues is addressed by the foundation's research agenda: resources, treatment and operations, distribution and storage, water quality and analysis, toxicology, economics, and management. The ultimate purpose of the coordinated effort is to assist water suppliers to provide the highest possible quality of water economically and reliably. The true benefits are realized when the results are implemented at the utility level. The foundation's trustees are pleased to offer this publication as a contribution toward that end.

Walter J. Bishop
Chair, Board of Trustees
Awwa Research Foundation

Robert C. Renner, P.E.
Executive Director
Awwa Research Foundation

FOREWORD

The Water Research Foundation is a nonprofit corporation that is dedicated to the implementation of a research effort to help utilities respond to regulatory requirements and traditional high-priority concerns of the industry. [illegible]

[illegible]

A broad spectrum of water supply issues is addressed by the Foundation's research agenda: resources, treatment and operations, distribution and storage, water quality and analysis, toxicology, economics, and management. The ultimate purpose of the coordinated effort is to assist water suppliers to provide the highest possible quality of water economically and reliably. The true benefits are realized when the results are implemented at the utility level. The Foundation's trustees are pleased to offer this publication as a contribution toward that end.

[illegible]
Chair, Board of Trustees
Water Research Foundation

[illegible]
Executive Director
Water Research Foundation

ACKNOWLEDGMENTS

The project team gratefully acknowledges the following for their support of this project:

Awwa Research Foundation
U.S. Environmental Protection Agency
American Water, New Jersey
Charleston Commissioners of Public Works, South Carolina
City of Cleveland, Ohio
City of Gastonia, North Carolina
City of San Antonio, Texas
City of Tampa, Florida
Columbus Water Works, Georgia
El Paso Water Utilities, Texas
Denver Water, Colorado
Minneapolis Water Works, Minnesota
New York City Department of Environmental Protection, New York
North Penn Water Authority, Lansdale, Pennsylvania
Portland Bureau of Water Works, Oregon
Richmond Department of Public Utilities, Virginia
Seattle Public Utilities, Washington

EXECUTIVE SUMMARY

This project was developed with two primary objectives: (1) to identify potential benefits to drinking water utilities from an environmental management system (EMS) and (2) to define steps necessary to develop a water utility sector–specific EMS model from which utilities could create their own EMS. As the project progressed, the project advisory committee (PAC) and utility workgroup determined that it would be most useful to develop this guidance using the existing International Organization for Standardization (ISO) 14001 voluntary standard as a model, on how to implement an EMS at a drinking water utility. This report reflects both the initial research into alternative EMS models and a "best practices" guidance based on the ISO 14001 framework.

Environmental management systems consist of linked sets of procedures used to systematically identify significant environmental issues within an organization and improve the management of these issues. Environmental management systems were first developed by large manufacturing companies to ensure that their environmental management actions were planned, implemented, tracked, and reviewed with consistent levels of quality. While a variety of EMS models have been developed, virtually all have core elements intended to drive the "Plan-Do-Check-Act" cycle of continual improvement:

- Planning
- Implementing those plans (do)
- Monitoring the results of plan implementation (check)
- Evaluating the monitoring results to support subsequent planning (act)

Drinking water utilities are facing increasingly numerous and complex challenges in managing environmental issues. An EMS is seen as a potentially effective tool to assist water utilities in structuring their environmental programs to achieve or exceed targeted levels of performance. Organizations using an EMS frequently report improved regulatory compliance, resource efficiency, improvements in their environmental performance, and improved relations with regulatory agencies. The fundamental environmental mission shared by water utilities provides a strong incentive to share and promote methods for improving the reliability and quality of water supply systems across the country.

Drinking water utilities implement management systems for a range of reasons, including:

- Achieve, maintain and document high levels of performance.
- Improve communication and knowledge management within the organization.
- Reduce environmental risks, and document the magnitude of the reduction.
- Improve credibility with the public and stakeholders with audits and improved communications.
- Improve response to unexpected events.
- Through the use of best management practices, including the use of EMS, avoid problems and reactive regulation.

This document was developed to assist water utilities interested in developing an EMS that will support and ensure continual improvement, increase thoroughness in compliance efforts, and demonstrate environmental excellence. It summarizes the experience and advice of leading private and public water utilities that have developed or are developing an EMS and concludes that a system structured around the ISO 14001 standard provides the greatest value for most utilities. The report is intended to provide a range of resources to utilities, including:

- Rationale for committing to the development of an EMS
- Explanation of EMS components and how they work
- How to start developing an EMS
- Suggestions for developing individual EMS elements
- Examples of EMS elements
- Factors to address that are unique to water utilities
- Recommendations for implementing an EMS

This report will be useful for utility managers interested in exploring the benefits of an EMS and valuable to utility staff charged with EMS development and implementation work.

The report includes results from a comprehensive review of the literature which provided a basis for evaluating EMS options for the drinking water industry, and also discusses results from a survey conducted to assess industry professionals' views about key issues facing the industry and how an EMS might help them manage these issues more effectively.

The main body of the report is divided into sections outlining EMS basics, followed by an examination of the challenges being faced by drinking water utilities, the activities that may result in environmental impacts that could be managed through an EMS, and a high-level view of the process for beginning the EMS development work. The report then discusses the specific elements of an EMS in greater detail, and provides guidance material for each element, including unique considerations for drinking water utilities. The report concludes with recommendations for future research.

The report also contains a number of examples or templates from utilities that have completed EMS development work. These materials are included to illustrate how EMS principles can be applied in practice and to provide ideas for EMS development, which can be readily adapted for use by other utilities.

The majority of EMS experience in utilities is currently with wastewater utilities; a relatively small number of drinking water agencies are using management systems. However, there are a number of common elements for developing a successful EMS that are important for both water and wastewater utilities such as:

- Commitment to EMS by top management is essential to develop, implement and maintain EMS. Management needs to demonstrate with words and actions why the organization is committing to a management system. Organizations that have recently managed change will be better prepared to manage the structured change involved in an EMS, and to integrate EMS functions into standard practices.
- Developing an EMS requires a designated coordinator to oversee and direct actions.
- Effective performance measures are essential for measuring progress and demonstrating the value of an EMS.

- Involvement by employees in the development of the EMS is essential to create "ownership" as well as getting accurate and relevant information on operations and goals.

CHAPTER 1
INTRODUCTION AND PROJECT BACKGROUND

PROJECT OBJECTIVES AND WHY AwwaRF CONDUCTED THIS PROJECT

The objectives of this project were to 1) identify the potential benefits of developing an environmental management system (EMS) specific to drinking water utilities, and 2) give these utilities a practical guide for doing so.

The reasons for initiating this project are perhaps best expressed by quoting from the AwwaRF request for proposals:

> Although water utilities have been viewed as clean facilities, they can have a number of impacts on source water, land use, and the environment. Water treatment plants consume significant amounts of energy and chemicals and simultaneously produce large waste streams. In addition, water treatment facilities are industrial facilities that store, use, and dispose of hazardous chemicals, which, if not properly handled, can lead to environmental degradation. Examples of these hazards include chlorine, fluoride, ammonia, petroleum products, acids, and bases. Smaller volumes of waste chemicals and sludge from laboratories and maintenance activities are also produced. Ensuring these products are transported, stored, used, and disposed of properly is a sound environmental practice and a sound business practice given the legal and regulatory repercussions of not dealing with them properly. Implementing an EMS that evaluates ongoing utility activities and includes performance measures, based on the basic Plan-Do-Check-Act approach that is the heart of an EMS, will help utilities demonstrate a commitment to environmental stewardship.

PROJECT TEAMS

Project Advisory Committee

The AwwaRF Project Advisory Committee (PAC) was composed as follows:

- S. April Richards, Environmental Engineer, U.S. Environmental Protection Agency (EPA)
- Jim Horne, Assistant to the Director, Office of Wastewater Management, EPA
- Belinda M. Minter, Environmental Manager, Department of Public Utilities, Water Division, Cleveland, Ohio
- David Lipsky, First Deputy Chief, Division of Water Quality Control, New York City Department of Environmental Protection

The PAC's role was to provide guidance to the project team, review interim products, and participate in the project workshops.

Utility Workgroup

A utility workgroup was formed from a representative set of small and large utilities, some with experience developing and implementing an EMS. The workgroup contributed to the development of this guidance through participation in two workshops and a draft review. The workgroup comprised representatives from the following:

- American Water, New Jersey
- Columbus Water Works, Georgia
- El Paso Water Utilities, Texas
- Denver Water, Colorado
- City of Gastonia, North Carolina
- Minneapolis Water Works, Minnesota
- North Penn Water Authority, Pennsylvania
- Portland Bureau of Water Works, Oregon
- Richmond Department of Public Utilities, Virginia
- Seattle Public Utilities, Washington
- City of Tampa, Florida

Project Team

- CH2M HILL: James Ollerenshaw, Paul Burnet, Bob Fuller, Mike Matichich, Joe Pringle, Alan Ispass, Elisa Speranza, and Becky Weig
- Ross & Associates: Rob Greenwood and Lori Ahouse
- International City/County Managers Association: Barbara Yuhas
- Center for Organizational and Technological Advancement, Virginia Tech: Bob Herbert

METHODOLOGY AND HOW THE GUIDANCE WAS DEVELOPED

EMS Literature and Initiatives Review

A literature review was conducted to identify and benchmark current approaches to environmental management systems and other relevant initiatives in order to provide a basis for developing and evaluating model EMS options for drinking water utilities. The review included other targeted programs and case examples of selected programs to illustrate possible approaches and benefits.

The initiatives were selected to provide broad coverage of management systems and other performance-improvement initiatives that are potentially useful for utilities in developing a drinking water EMS. The initiatives include internationally accepted, voluntary management system standards; customized EMS programs and guidance from other industries, including wastewater utilities and local governments; best practices specific to the drinking water industry and public utilities; recognition programs for quality management; product labeling programs; and international performance improvement and certification programs.

These programs were examined for concepts and components that potentially could be modeled or integrated into an EMS for a drinking water utility. The literature review materials highlight the experiences of drinking water utilities, wastewater utilities, and other organizations in developing EMS programs and other management initiatives, providing valuable experience for EMS development for water utilities.

Included in Appendix B are (1) a reference sheet providing an at-a-glance view of the attributes and characteristics of each initiative included in the literature review, (2) an annotated bibliography with reference materials for each of the initiatives, (3) a comparison matrix describing and comparing the initiatives' key attributes and components, and (4) a subset of seven initiatives selected for more in-depth examples illustrating successes and lessons learned.

Survey

The project team conducted a web-based survey to assess current views about key issues facing drinking water utilities, such as regulatory compliance, product and service quality, and environmental performance, and to evaluate how an EMS might help manage these issues more effectively. The survey included individuals working in the drinking water industry, as well as others who had direct experience in implementing environmental management systems for public utilities. They were identified from a variety of sources, including contacts that the project team and its extended contact group identified as individuals potentially interested in EMS issues or likely to have been involved in activities that would be related to EMS programs.

The survey was not designed to produce statistically significant results, but to provide additional background to the project team and the utility workgroup on drinking water utilities' unique needs, priorities, and challenges. Given the very limited information related to EMS programs within the water industry, which is primarily limited to the case examples of the few utilities that have fully implemented EMS programs, the survey provided valuable information for this project.

In total, 343 people took the survey. Additional information and complete survey results are in Appendix C.

Project Workshops

Valuable information for the project was gained through two workshops held in May and November 2004. These workshops included project team members, the utility workgroup, the project advisory committee, and invited guests, including representatives from the Charleston Public Utility Commissioners of Charleston, South Carolina, and the City of San Antonio, Texas). The first workshop clarified the issues that are of prime importance to drinking water utilities, and also explored the EMS structure that was favored by the workgroup. The second workshop focused on the system elements and their relationship to drinking water utilities.

Additional information on the workshops is included in Appendix D.

CHAPTER 2
INTRODUCTION TO ENVIRONMENTAL MANAGEMENT SYSTEMS

WHAT IS AN ENVIRONMENTAL MANAGEMENT SYSTEM?

An EMS is a framework, consisting of a defined set of management procedures, for improving the operations and results relating to environmental impacts of an organization. This framework uses interconnected procedures to ensure that best management practices are consistently defined and used within any part of an organization to reduce environmental impacts and continually improve organizational performance. The EMS approach is based on the principles of total quality management, which were initially developed by the private sector to achieve higher and more-consistent product quality. Many prominent companies in the United States have embraced environmental management systems, including IBM, Ford, Colgate Palmolive, and Lockheed Martin, and a number of public organizations at the local, state and federal levels are also successfully using management systems.

Key functions of an EMS include:

- Defining potential environmental impacts and setting priorities based on these impacts
- Setting goals for improving environmental performance, and defining methods for measuring progress toward these goals
- Defining procedures to ensure that performance will be maintained or improved
- Defining roles and responsibilities to consistently apply procedures
- Checking actual performance and making corrections accordingly
- Communicating performance to management and
- Defining new goals to ensure continual improvement.

The cornerstone of an EMS is its regular continual improvement cycle of planning, implementation, performance monitoring, and review. This "Plan-Do-Check-Act" cycle of an EMS provides the procedural "flywheel" for improving environmental performance over time. (See Figure 2.1.) Some utilities have processes in place that cover the planning and implementation part of the cycle—"plan" and "do"—but they may lack a defined system to regularly monitor performance and make management decisions based on real outcomes—the "check" and "act" parts of the cycle.

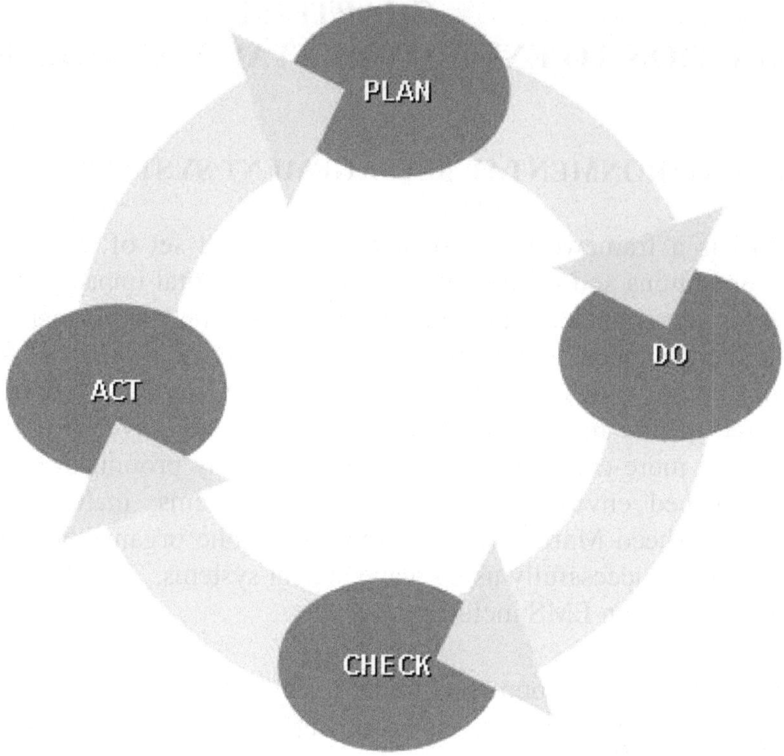

Figure 2.1 EMS Flywheel

Typically, an EMS is used to support continual improvement of activities relevant to environmental performance by helping an organization identify and act on opportunities for improvements in the following areas:

- Environmental performance and pollution prevention
- Product quality and best management practices
- Communication and credibility with stakeholders
- Regulatory compliance
- Risk mitigation and preparedness
- Knowledge management and succession planning
- Competitiveness and cost savings
- Security

In addition, an EMS framework can be adapted to support the needs, priorities, and circumstances of specific industries or individual organizations. Water utilities, sharing many common challenges, have the opportunity to collaborate in developing tailored resources for the industry. This document was developed as an initial tool for this purpose.

A growing number of public organizations, including drinking water and wastewater utilities, are developing and applying EMS programs to ensure consistent planning, performance and continual improvement. An EMS is structured to help guide an organization through a regular process of ensuring that best practices are defined and used. While simple in concept, this "say what you do and do what you say" process is structured around component pieces that collectively provide a robust program that can be checked, or audited, to provide documentation that best practices are, in fact, being used. The net result is a management structure for helping ensure that key tasks are performed properly and on time in a way that enhances overall performance and builds credibility with management and stakeholders.

Drinking water utilities may see significant benefit from using an EMS, some general and some specific to the circumstances of each agency. Among likely and potential benefits, an EMS:

- provides a recognized, auditable structure for ensuring that best practices are consistently used
- ensures that the chance for mistakes and errors will be reduced, and that when they occur corrective actions will prevent recurrence
- provides the transparency expected of many public agencies; third party audits also provide a high degree of credibility to outside observers
- ensures that new regulations are identified early and that roles and responsibilities, training, monitoring requirement and other factors are addressed
- ensures clear descriptions of procedures and best practices to ensure smooth succession planning and training of new staff
- assists high-performing drinking water utilities to better document and tell their story to management, elected officials, regulators, interested parties, the media and the public.

COMPONENTS OF AN EMS AND HOW THEY WORK

Although the ISO 14001 standard has been established, many organizations are "customizing" environmental management systems to meet their unique needs and circumstances. There are a number of common components to any EMS. These components, or "elements," are linked together in the plan-do-check-act cycle of continual improvement described above. Elements are described in detail in Chapter 5 of this report. The relationships among elements are shown in Figure 2.2. The core elements of any EMS are described below:

Environmental Policy Statement

A policy statement provides high-level principles that declare the organization's values and commitments in areas such as environmental performance, pollution prevention, compliance, continual improvement, communication, and use of best-management practices.

Legal and Other Requirements

Identifies and tracks regulatory requirements and other environmental commitments that an organization has made and which must be supported by the EMS.

Evaluation of Environmental Impacts and Opportunities for Improvement

An evaluation of environmental impacts provides a structured analysis and prioritization of how the organization's activities impact the environment, what activities must be managed to prevent environmental impacts, and opportunities for performance improvement.

Goals, Objectives, Targets, and Management Programs

This linked set of elements describes mechanisms used to achieve desired performance improvements, including goal statements, quantitative targets for improvement, and action plans for meeting desired improvements.

Roles, Responsibilities, and Competency

These elements identify the positions, duties, and training of staff members whose roles are related to management programs and whose day-to-day work responsibilities include those that have the potential to affect the environment or are otherwise related to environmental performance.

Communication

This element defines procedures for communication of information about the organization's environmental performance, achievements, compliance, and activities related to the EMS to internal and external stakeholders. The communication element can include mechanisms for responding to requests for information as well as receiving, considering, and responding to input from external stakeholders.

Documentation, Document Control, and Recordkeeping

These procedures describes how an organization's EMS "works" (i.e., how the organization identifies environmental impacts, how it sets objectives and targets, how it audits the system, etc.) Documentation control includes procedures to manage the creation, revision, availability, and approval of documents related to the EMS. This element also covers recordkeeping, which includes keeping track of EMS-related activities such as records of completed actions , records of performance against objectives in targets, communication records, and training records.

Operational Control

This element describes procedures and management practices used to control activities that can impact the environment.

Corrective and Preventive Action

This element defines a structured process for identifying and correcting root causes of incidents, mistakes, and problems that have the potential to lead to non-compliance with applicable regulatory requirements or significant impacts to the environment.

Monitoring and Measurement

Effective measurement is critical to managing; this element provides standard practices and procedures for measuring and monitoring activities. The element includes developing performance indicators that allow an organization to effectively track environmental performance and as progress towards identified targets. In addition, this element includes a process for evaluating and reporting these measurements to decision-makers and staff.

EMS Auditing and Management Review

These procedures evaluate on a regular basis the effectiveness of the EMS, progress towards objectives and targets, performance indicators, problems, and achievements.

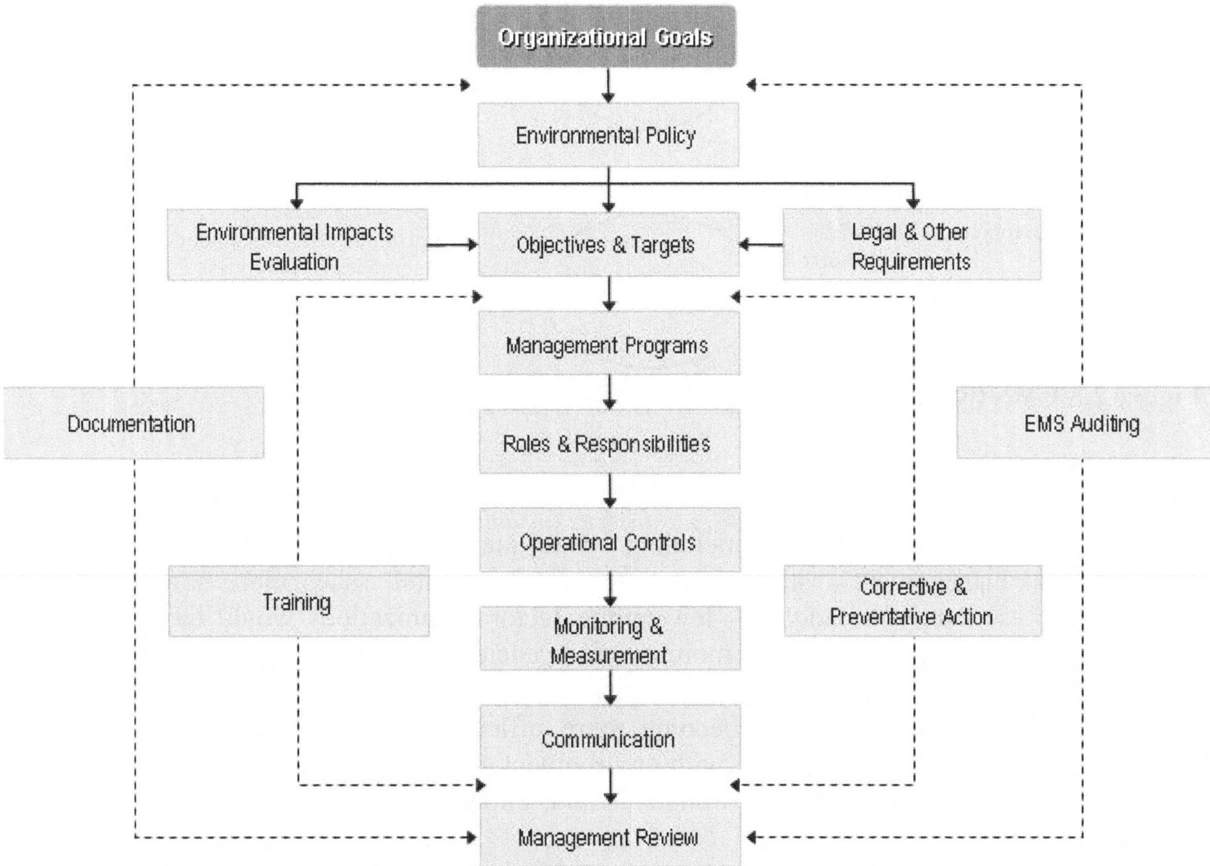

Figure 2.2 Relationships among EMS elements

Figure 2.3 shows the EMS elements arranged in the Plan-Do-Check-Act flywheel, illustrating how they fit into the development and implementation cycle:

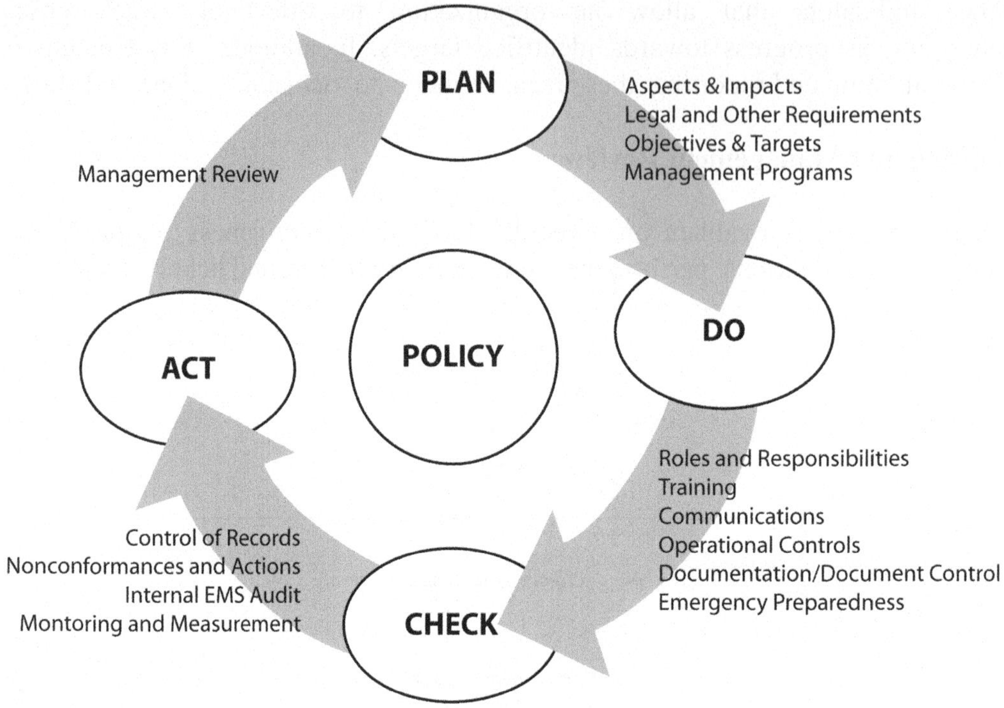

Figure 2.3 EMS flywheel with system elements

EMS IN THE PUBLIC SECTOR

Local government agencies including wastewater utilities are increasingly turning to environmental management systems to manage environmental issues more strategically and proactively. Less than a decade ago, few public sector organizations would have considered implementing an EMS. However, as more public agencies began transforming themselves into competitive, business-minded enterprises, they looked to the private sector for tools and approaches that would help them become more efficient and effective at achieving their missions. For many, especially those with environmental stewardship responsibilities, an EMS fits well with existing programs and initiatives and, importantly, helps organizations fill some key gaps.

In the United States today approximately 200 public agencies have or are developing EMS programs, including almost 100 wastewater treatment facilities, and a handful of drinking water facilities. Local governments also have developed environmental management systems for solid waste, transportation, and parks and recreation operations. The Office of Water at U.S. EPA has been actively promoting EMS adoption by local governments, including water and wastewater utilities, since 1997. EPA has assisted over 30 agencies and has also supported a national Clearinghouse of EMS information for local governments at www.peercenter.net. This Clearinghouse contains case studies, guides, and much more information geared to the needs of local governments, including a companion EMS guide for wastewater utilities entitled *Achieving Environmental Excellence: An EMS Handbook for Wastewater Utilities.*

WHY DOES A UTILITY IMPLEMENT AN EMS?

There are a variety of reasons a public utility may choose to commit to using an EMS. While the primary factors may vary among utilities, depending on size, political structure, public perceptions and a host of other factors, common considerations include:

- Achieve and Demonstrate High Performance
  - Produce consistent, high-quality products and services
  - Meet or exceed regulatory requirements
  - Avoid mistakes that can result in environmental impacts, compliance problems, and negative publicity
  - Track and report achievements and performance
- Improve Communication and Knowledge Management
  - Improve internal communication and decision-making, and integrate artificially isolated work areas, such as operations and maintenance
  - Define and implement critical procedures
  - Engage critics and supporters more effectively
  - Improve accessibility of documents and information
  - Clarify responsibility and accountability, authority, training and expectations for staff
  - Move away from a "someone-else's-problem" mentality
- Reduce Environmental Risk
  - Avoid mistakes that can result in environmental impacts, compliance problems, and negative publicity
  - Move away from a "someone-else's-problem" mentality
  - Improve environmental emergency preparedness and response plans
- Improve Public Image
  - Receive certification or "seal of approval"
  - Improve credibility with public
  - Be perceived as an industry leader
  - Participate in citywide EMS initiatives and environmental management systems for local governments
  - Improve competitiveness
  - Better manage processes and products
- Improve Response to Unexpected Events or Compliance Problems.
  - Respond quickly and effectively using defined procedures
  - Strengthen ties with emergency responders at local, county, state and federal levels
  - Test response procedures to ensure effective actions in the event of an event
  - Use root-cause analysis and actions plans to document that causative factors are addressed to prevent recurrence
  - By effectively addressing potential or actual problems, avoid reaction from the public in the form of reactive regulation, as illustrated in Figure 2.4.

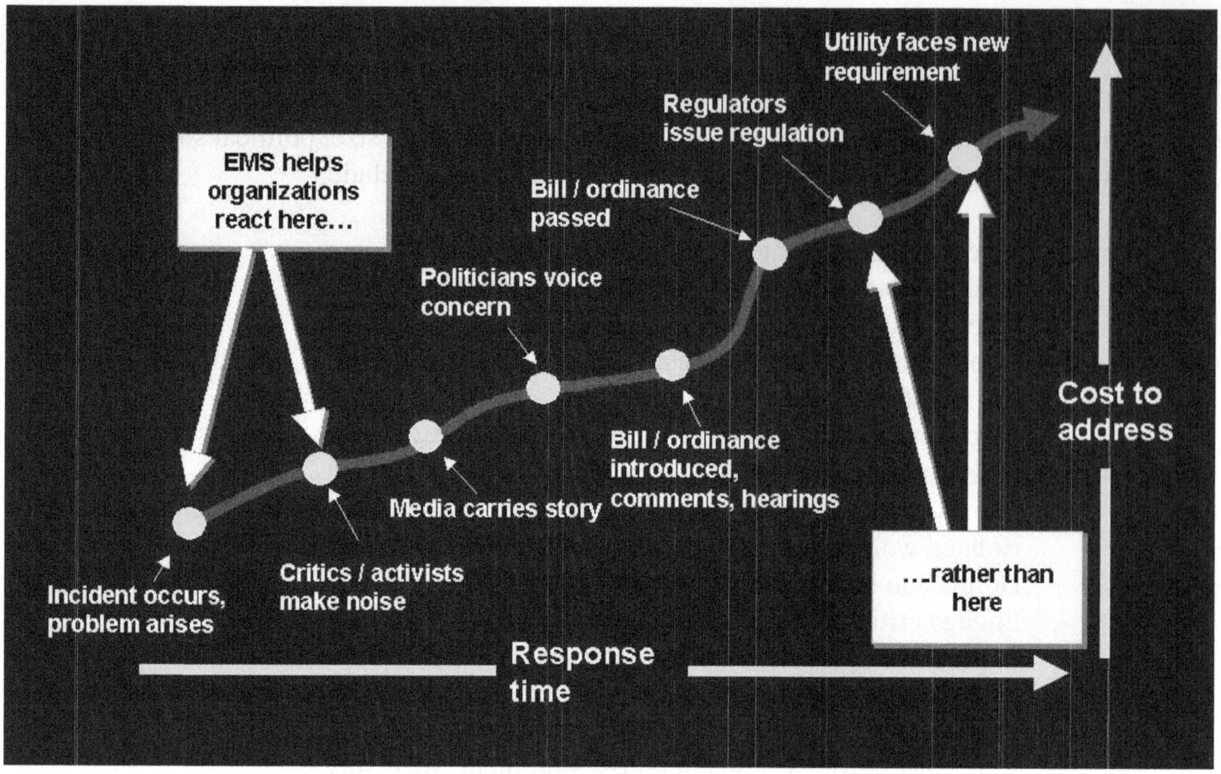

Figure 2.4 How EMS helps organizations react faster and respond better

COMMON EMS MODELS AND INITIATIVES

ISO 14001

The International Organization for Standardization (ISO) has developed a voluntary EMS standard (14001) that has been adopted by thousands of organizations globally and in the United States. The ISO 14001 EMS includes 17 elements that comprise the plan-do-check-act framework. While other EMS models are in use, the universal applicability and recognition of the ISO 14001 standard provide a highly credible framework for water utilities interested in developing their own EMS. Guidance on developing an EMS at a drinking water utility, provided in this document, is based on the 17 elements of the ISO 14001 EMS standard which are:

1. Environmental policy
2. Document control
3. Environmental aspects
4. Operational control
5. Legal and other requirements
6. Emergency preparedness and response
7. Objectives and targets
8. Monitoring and measurement
9. Environmental management program(s)
10. Nonconformance and corrective and preventive action

11. Structure and responsibility
12. Records
13. Training, awareness, and competence
14. EMS audit
15. Communication
16. Management review
17. EMS documentation

The "EMS Integration Project" (www.wef.org)

In 2002 the Water Environment Federation (WEF), the National Association of Clean Water Agencies (NACWA, formerly the Association of Metropolitan Sewerage Agencies), and the EPA initiated a project to develop a template to assist utility managers understand and integrate environmental management systems and other management programs, initiatives, and tools into an accepted, universal management system framework, using the plan-do-check-act cycle of a continual improvement management system.

National Biosolids Partnership (www.biosolids.org)

The National Biosolids Partnership (NBP) was founded in 1997 by NACWA, WEF, and the EPA, with a goal of advancing environmentally sound and accepted management practices. Starting with 13 charter pilot agencies, the NBP program now includes over 80 utilities that are developing environmental management systems for their biosolids operations with plans for more. The NBP has developed an EMS focused on the biosolids (digested solids resulting from sewage treatment) production chain. The NBP EMS model is based on the ISO 14001 standard but tailored to the needs of wastewater treatment plant operations and, in particular, pubic relations regarding biosolids and end uses of the material. Completion of an independent third party EMS audit is also a required part of the NBP program.

The experiences of agencies currently developing an EMS as part of the NBP EMS demonstration program offer some insight on implementing an EMS in a utility that has a core environmental mission.

EPA's EMS Initiatives for Local Governments (www.peercenter.net)

The U.S. EPA Office of Water has supported a total of 32 local government agencies developing environmental management systems under this program. These initiatives have helped demonstrate the significant benefits of EMS in the public sector, and have established a foundation to further promote EMS adoption in public agencies. Case studies on each of these agencies are also available.

Public Entity Environmental Management System Resource Center

EPA, working with the Global Environment and Technology Foundation (GTEF), has also established the Public Entity Environmental Management System Resource (PEER) Center. The PEER Center has two major components. The first is a virtual clearinghouse for local, county, and state governments that are considering implementing or have implemented an EMS and want to access the knowledge and field experience of other public entities that have done so. The second component consists of eleven EMS Local Resource Centers (LRCs) around the country. These Centers are intended to promote EMS competence and encourage government-to-government sharing and mentoring among public sector organizations. Each PEER Center seeks to provide technical expertise, field-tested tools, and support for EMS implementation to interested agencies.

CHAPTER 3
HOW AN EMS CAN HELP DRINKING WATER UTILITIES IMPROVE PERFORMANCE AND PRODUCT QUALITY

AN EMS FOR DRINKING WATER UTILITIES

Drinking water utilities perform a critical mission every day to deliver high-quality products and services. This mission requires adapting to changing regulations and community needs, reporting on performance, and above all, safeguarding human health. Drinking water utilities also face pressure to be more accountable on environmental issues and more accessible to a wide range of stakeholders.

Meeting these responsibilities is challenging in the face of flat or declining net revenue. It is further complicated by priorities such as the need to enhance security and to maintain and replace aging infrastructure. Environmental management systems offer drinking water utility managers a strategic approach to meeting these challenges and improving organizational performance.

Drivers and Benefits

The drinking water utility workgroup identified the following common drivers and benefits for implementing an EMS:

- Improved efficiency of operations
- Reduced noncompliance
- Desire to be "best in class"
- Improved training and awareness
- Use of a fundamental best practice
- Public credibility, which makes it easier to sell tough decisions such as rate increases
- Senior management impetus for a management system
- Need to prevent, not just respond to issues.
- Continuity of operations and institutional knowledge

Challenges Facing Drinking Water Utilities

Participants in the first project workshop engaged in an exercise to brainstorm the challenges facing utilities and then group the challenges under various categories. The results of this exercise, and how an EMS can help overcome these challenges, are shown in Table 3.1:

Table 3.1
Drinking Water Utility Challenges and EMS Benefits

Challenge	EMS Benefits
Performance Measures—Challenges include issues such as how to better measure utility effectiveness, and how to identify the appropriate indicators. This includes consideration of the implementation of performance measurement mechanisms and tools—testing procedures, audits, quality checks, etc.	Goals and objectives defined by the EMS will help define clear, measurable targets. The structured process for defining goals helps ensure that goals are measurable and that data to measure progress toward those goals can and will be collected. Development of measures is usually conducted more consistently under an EMS, allowing preparation of performance reports that accurately convey the overall performance of the utility.
External Communication—Effective communication with external stakeholders such as the general public and other customers is an ongoing challenge for drinking water utilities. In addition to providing environmental education to the public, communication goals also include justifying rate increases, ensuring the public's trust and confidence in the water supply, and receiving effective coverage in local media. A related issue is facility security and how to balance onsite tours and educational events with enhanced security measures.	Clearly defined goals, effective performance measures, annual performance reports and audits—all parts of an EMS—can help communicate to the public the level of quality of a utility's operation. These benefits becomes even greater over time, when the cycle of goal setting and goal achievement, and successful audit cycles, can be clearly shown.
Resource Constraints—A continuing challenge for the industry is how to do more with less, remain competitive, and find the resources to maintain excellent operations with a highly trained work force.	Effective communication with the public if often a key factor in securing funding necessary for infrastructure, security, testing or operational improvements. An EMS helps enhance the credibility of decision-making by communicating how goals and priorities are set, documenting efficiency, and explaining regulatory drivers and desired actions (for example, emergency response drills) that may not be currently in place.
Work Force Effectiveness—Maintaining and improving work force effectiveness confronts many utilities. This is an issue both of ensuring skill and knowledge transfer to replacement staff and maintaining and recruiting excellent staff. Other issues include staff acceptance of new technology, cross-training, improving productivity and motivation.	An EMS ensures that operational controls and standard operation procedures are prepared for all aspects of utility operation. This helps ensure that knowledge transfer is occurring, ensures that best practices are used consistently, ensures that appropriate training for new and existing staff are conducted, promotes efficient succession procedures and helps enhance communication among staff and management. The resulting work environment can be a factor in attracting and retaining the best work force.

(continued)

Table 3.1 (Continued)
Drinking Water Utility Challenges and EMS Benefits

Challenge	EMS Benefits
Risk Management—Risk management and security issues continue to be challenges. Utilities can struggle with risk avoidance and planning—appropriately tracking regulations and compliance—and with risk prioritization.	An EMS requires clear procedures for tracking applicable rules and regulations, and assigns roles and responsibilities to ensure these tasks are conducted. Similarly, if risk management or security enhancement tasks are identified as high priority goals, the EMS procedures will help ensure that progress is made toward achieving those goals.
Water Resource Planning—Ensuring future water supply to meet increased and managing unaccounted for water are frequent ongoing challenges.	As noted with Risk Management, key planning and problem resolution issues can be included as EMS goals, ensuring that staffing and measurement of progress toward goal achievement is conducted.
Strategic Planning—Lack of or inadequate comprehensive strategic planning is identified as a common challenge for utilities. Establishing a clear vision for the strategic plan and having the right people to provide the vision is an associated challenge. Identifying and incorporating community values and setting appropriate environmental goals are also important. Having established a strategic plan, Maintaining management focus during implementation of the strategic plan is also crucial.	An environmental policy is an essential element for an EMS, and establishing that policy allows goals and objectives to be developed that are consistent with that policy. The continual improvement "flywheel" of an EMS also ensures that change will be ongoing, and helps support longer-term strategic planning. An EMS can be structured to have specific community involvement components as well.
Operations Resource Planning—Planning and obtaining necessary resources for replacement or upgrade of operations infrastructure can be challenging. An effective preventive maintenance program can reduce the severity of this challenge.	Preventative maintenance procedures are considered an essential part of operational controls and are required by an EMS. Tasks related to maintaining or upgrading infrastructure or operations can be managed through goal setting, and the need for actions can be shown using data obtained through measuring and monitoring actions.
Cost Management—Management of costs and controlling rates is a common utility challenge, and investment needs must be understood and prioritized. New and improved technology is a part of this evaluation.	The need for and evaluation of technology improvements can be supported by the process of goal setting, building trust through communicating with stakeholders and employees and demonstrating excellence via the audit process, and ensuring that plans and strategies are clearly linked with goals and objectives.

(continued)

Table 3.1 (Continued)
Drinking Water Utility Challenges and EMS Benefits

Challenge	EMS Benefits
Culture Change—A challenge with implementing new systems or programs such as an EMS is managing the associated culture change in the organization. In some cases, operators or other employees are hesitant to accept new technology or procedures, and some work areas or departments are hesitant to acknowledge an environmental connection to their work.	An EMS requires a clear definition of roles and responsibilities, training requirements, and documentation of SOPs. The audit process reinforces the need to follow written procedures. The process of developing an EMS can help improve staff skills in managing culture change, and the EMS structure helps ensure that changes are implemented consistently.
Knowledge Management—Knowledge management is an ongoing challenge for many utilities. Effective systems for data management and documentation (including software) are needed. Capturing and acting on staff ideas and suggestions and systematically transferring staff knowledge are related challenges.	An EMS is structured to ensure that SOPs are documented and that documents and records are managed effectively. If developed in an electronic format, and EMS ensures (and encourages) access to current resources by all staff. The goal setting process, and any nonconformance corrective actions, creates opportunities for soliciting, obtaining and using staff ideas and solutions.

An EMS is a structured set of procedures that can help utilities address many of these challenges. Drinking water agencies that implement an EMS need to develop and document procedures, and implement training programs, which directly or indirectly help to address many of the challenges listed above. The elements of an EMS systematically help promote improved performance measurement, communication, training, change management, and knowledge management.

ENVIRONMENTAL IMPACTS IN THE DRINKING WATER INDUSTRY

Drinking water utilities produce a high-quality product that is critical to our health and well being. Sometimes this service is accomplished by treating water that would otherwise be undrinkable or unsuitable for other uses, thus quantifiably improving human health and the environment. However, if not managed effectively, the processes and procedures necessary to produce clean, safe drinking water have the potential to impact water, land, and air resources in negative ways. An EMS can be a critical tool for utilities working to reduce or eliminate these impacts.

Similarities among drinking water utility operations allow potential environmental impacts to be listed for the industry; actual potential impacts can then be selected by individual utilities. The following list of activities, including some potential associated environmental impacts, was developed at the first workshop:

- Operation of treatment processes and equipment (energy use, noise, potential habitat impacts)
- Use of treatment chemicals (disposal, spills, transportation impacts)
- Use, storage, and disposal of hazardous chemicals (disposal, spills, transportation impacts)
- Laboratory operations (disposal, spills, transportation impacts)
- Residuals management and disposal (disposal, spills, transportation impacts)
- Maintenance (solvent use/chemical)
- Used parts disposal (disposal, replacement manufacturing)
- Landscaping and grounds maintenance (pesticide use, energy use, invasive species, water use, stormwater)
- Water main construction repair and replacement (stormwater, energy, disruption of transportation, noise)
- Water line flushing (surface water impacts)
- Transportation and fleet maintenance (energy use, solvent/chemical use)
- Facility construction (habitat loss, stormwater, material supply, energy)
- Water conservation programs (stormwater,
- Source water protection and watershed management (habitat, surface water)
- Water withdrawals (aquifer/surface water)
- Facility security (chemical releases, protection of human health)

Unlike manufacturers and many other organizations that have implemented an EMS, the "environmental footprint" of drinking water utilities is not limited to undesirable environmental impacts that are byproducts of their operations. Drinking water utilities' primary mission is inherently environmentally related because they produce and deliver an environmental "product" to the public. Everything the utilities do, including business processes, procedures, and activities, supports this fundamental environmental mission.

For this reason, when the environmental and public health impacts of water utilities do become visible, they tend to attract a great deal of attention—as certainly has been the case with outbreaks of *E. coli* in public water systems, as evidenced by the Walkerton case. Interestingly, recommendations from the official government inquiry that followed this incident included many of the key elements of an EMS.

While environmental management systems have been employed in response to a utility's failure to protect human health, the proactive use of an EMS to improve compliance and promote continual performance improvement is generally a more attractive approach for utilities, and is consistent with the leadership of EPA and many States in exploring and piloting environmental management systems "as a valuable tool to help organizations improve their environmental performance, increase the use of pollution prevention, and improve compliance" (EPA 2005).

Drinking water utilities and other public agencies that have implemented an EMS have found benefits that have included:

- Cost savings from reduced energy use, waste disposal and other sources
- Improved communication with staff, management, elected officials and the public
- More consistent practices, with fewer compliance or "near miss" incidents
- Improved trust and relationships with regulatory agencies

- Better succession planning
- More proactive problem solving by staff

CHAPTER 4
CRITICAL FACTORS FOR THE SUCCESS OR FAILURE OF AN EMS

KEY FACTORS FOR EMS SUCCESS

Top Management Commitment and Involvement

All guidance documents on EMS implementation, as well as practical experience, stress the importance of top management support and commitment to the EMS effort in their organizations. This also includes ongoing involvement with the EMS process, at a minimum through the management review process but ideally at a greater level.

Equally important as the initial commitment and support of top management is the demonstration of ongoing support. As the EPA wastewater handbook states "visibility, commitment, and involvement are the #1 keys to success" (EPA 2004). This can take many forms, according to the organization and the personalities of the top managers, but "regular and often" demonstration of support is essential.

Another aspect to management support is making sure that staff has sufficient time and resources available to successfully develop the system. This may mean temporary or permanent reallocation of duties or adjusting schedules for individuals or projects. Developing an EMS must be managed as a top priority for the organization if the effort is to be successful.

Designated EMS Coordinator or Champion

The EMS Coordinator for an organization is sometimes referred to as the "EMS Champion," and this is perhaps a better title. The EMS Coordinator has the overall responsibility and authority for developing and implementing the EMS. Selecting the right individual for this key role can make the difference between success and failure. The individual selected for this role needs to possess some specific skills and abilities, such as

- Ability to be respected by and communicate with all levels of staff
- Management and team leadership skills
- Scientific/environmental background
- Understanding of the organization (operations, business, etc.)

Feedback on Performance

Performance measures, especially when used for reporting to management and communicating with staff and stakeholders, are a critical part of an effective EMS. They are linked with goals, objectives, and action plans to create a clear pathway toward completion of priority tasks. Measures that are developed with input from employees will be accepted and used more readily that those that are imposed by management.

Organizational Relevance

Necessary changes in the way operations are conducted, recorded or managed may be resisted as unnecessary or ill-advised unless there is a clear reason or context for change. An EMS can help provide the driver for change, and thereby improve staff acceptance of the change process. This process may be reinforced by the audit component of an EMS. Knowing that a third-party auditor will be checking to ensure that procedures are established and followed (and will be reporting findings to management) can provide motivation and commitment to the changes an EMS may bring.

Implementing EMS requires organizational change. Organizations have success when they have a culture that is accepting of change. One way to help overcome resistance to change is to involve employees and management often in EMS planning and implementation, assign roles to individuals and teams to encourage ownership, set deadlines to ensure tasks are completed, and celebrate the accomplishment of significant milestones.

Employee Involvement

Employee involvement in EMS planning and implementation can help ensure that the system will become part of everyday business and not just a series of documents on the shelf. Employees can be involved in a variety of ways, including (1) identifying objectives and targets; (2) developing management programs; (3) establishing standard operating, monitoring, and measuring procedures; (4) participating in EMS training programs; and (5) implementing communications programs.

It is also key that employees understand their roles and responsibilities with respect to the EMS and for managing for environmental impacts. This is supported in the EMS elements covering roles and responsibilities, training, and standard operating procedures. The EMS structure can help provide accountability for tasks, both in development and implementation phases.

EMS Readiness

A key to successful EMS implementation is an assessment of the "readiness" of the organization to undertake this new initiative. EMS readiness involves an organization's ability to make changes required by an EMS structure, and was a topic of discussion at the second project workshop. The following were identified as important issues to be considered prior to starting EMS implementation:

- Obtaining upper- and middle-management support
- Ensuring participation and commitment from middle managers
- Ability to incorporate EMS fundamentals into the utility's strategic planning activities
- Ensuring that EMS elements are aligned with other business goals
- Demonstrating the need for change by emphasizing that investment now will provide cost and efficiency benefits later
- Defining the scale of operations to be covered by the EMS—this definition of the "fenceline" should consider whether one area of the utility may be more "ready" for change than another

- Understanding and being prepared for the culture change that will come as EMS is embraced by the organization

In order to provide further insight into an organization's readiness for EMS development and use, the concept of the organizational change equation was defined (NACWA and AMWA 1999). This definition states that need for change (urgency), the direction of change (vision) and the specific actions required (solution), occurring over a fixed period of time (time) must be greater than the anticipated resistance to change:

$$\frac{\text{Urgency} \times \text{Vision} \times \text{Solution}}{\text{Time}} > \text{Resistance}$$

Recognizing that there is a natural resistance to change, achieving a successful implementation of the new process of an EMS, the organization must overcome this resistance by developing and sustaining urgency for adopting new practices, focusing on a vision for improvement, and developing the solution for achieving that improvement, in this case through the implementation of an EMS. The less time it takes for a critical mass of organizational members to achieve the urgency, vision, and solution, the greater the potential to overcome the resistance and successfully accomplish the change.

Workshop participants identified several notions for each of the terms (except for time) in the equation, which may be of assistance to those EMS champions in a utility. Those are listed in Table 4.1.

Table 4.1
EMS Readiness Factors

Examples of Issues That May Create Urgency for Developing an EMS	Examples of Issues That May Create a Vision to Support EMS Development	Examples of How an EMS Can Be Shown as a Solution	Examples of Resistance to Be Overcome in Developing an EMS
• Improve organizational understanding of environmental impacts, legal and regulatory requirements • Reduce potential for a significant environmental impact • Mandate or directive from utility's governing body or CEO • Help defend projects, budgets, capital improvement plans • Reduce security risks and improve emergency response • Comply with and integrate new regulations into processes and procedures • Avoid regulatory enforcement and administrative penalties that creates a loss of control • Reverse perception that utility doesn't know what it's doing	• Improve community perception of utility and build confidence in water supply • Achieving EMS certification provides evidence of quality organization and environmental stewardship • Linkage between operations, capital planning and strategic business plan • Improve performance tracking and consolidate reporting • Improve efficiency of operation and maintenance through improved operational controls and proactive vs. reactive maintenance • Improve bond rating • Enhance organizational pride • EMS helps avoid "dropping the ball"	• Use EMS as a vehicle to accomplish an existing strategic goal(s) • Begin communication to all employees about EMS early • Possibly use a different name for EMS (e.g., "Earthwise") • Take a retroactive view of how an EMS could have saved money and loss of goodwill • Offer an EMS as a supplemental environmental project in lieu of enforcement action • Work with outside organizations to gain political momentum • Educate key stakeholders (e.g.: financial analyst, bond counsel, legal counsel) • Make the "why" personal by including participation as part of performance reviews • Lay out all of the pieces and identify the benefits for each piece • Publicize and celebrate easy wins as they develop	• Traditional organizational culture with silos and territories • Many other compelling priorities • Understaffing; insufficient personnel resources; insufficient time • Lackadaisical staff; What's in it for me? • Cost constraints; insufficient funds to pursue; no apparent return on investment • Already "do it;" no need for a formal and costly program • Concern for legal implications • Lack of information • Bargaining unit agreements and work rules • Lack of leadership; no champion, or champion lacks position of leverage • Inertia of middle managers that have spent entire career at utility • Lack of access to training

BARRIERS TO EMS SUCCESS

Developing and implementing an EMS involves managing change and requires a commitment of time and resources to make the change effort a priority. In a survey conducted by CH2M HILL for NBP, commitment and organizational barriers were cited most often as impediments to successful EMS development and implementation. Survey results are shown in Figure 4.1. These results are also consistent with the findings from many other EMS-based programs, including EPA's initiatives with local governments.

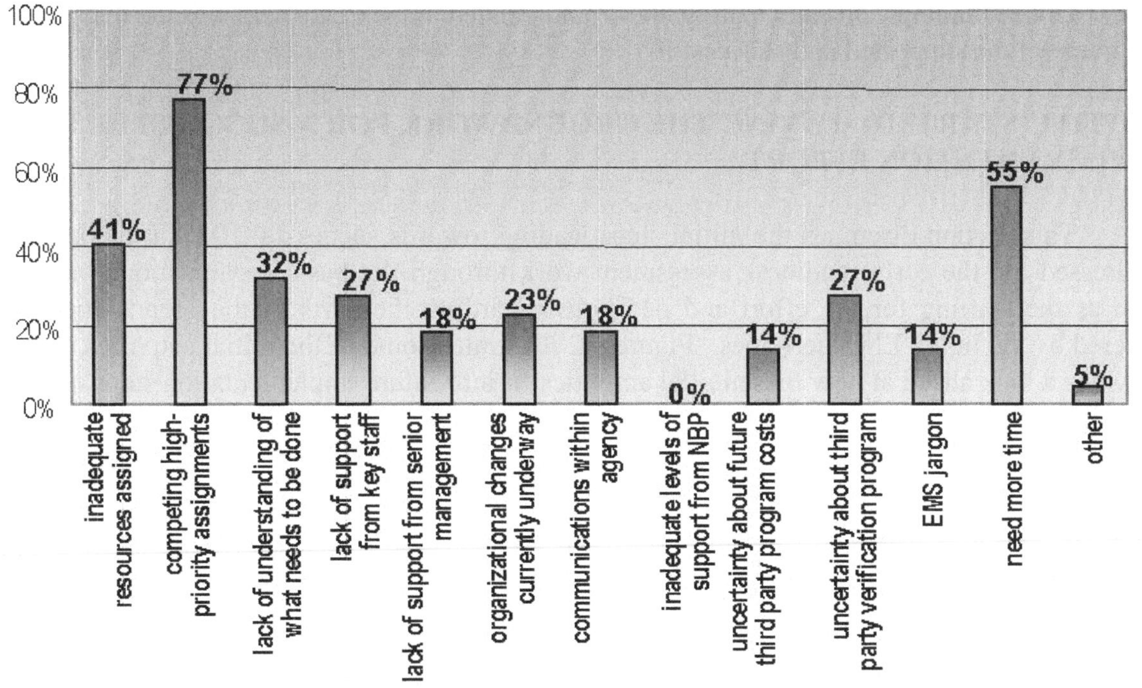

Figure 4.1 Impediments to successful EMS development and implementation

An examination of the factors responsible implementation problems can provide valuable lessons for organizations considering an EMS program. Common implementation challenges faced by public utilities include:

- Lack of commitment and competing priorities—the EMS may be viewed as just one more thing to do, another task "added to the plate," without a clear sense of relative importance
- Turf protection or "not my problem" managers or staff may not see an EMS, or components of an EMS, as necessary or relevant to their job, or may see it as empowering staff to change procedures without their supervisory approval
- Lack of understanding/jargon overload—an EMS has its own jargon which can be a barrier to understanding and acceptance of what components are and how they help support good practices

- Resistance to change—an EMS does require change, and some individuals or organizations are more resistant to change than others
- "Just get it over with" syndrome—managers or staff may go through the motions without fully understanding or embracing the changes required by an EMS, resulting in the EMS malfunctioning after it has been developed
- Emphasis on the EMS manual instead of real improvement—managers or staff may see developing an EMS as a single, one-time task rather than a new way of managing operations, and may be surprised or confused when the program continues after the EMS is developed.

Understanding and anticipating these and related issues can help ensure that an EMS program will be supported and successful.

GETTING STARTED—LAYING THE GROUNDWORK FOR A SUCCESSFUL EMS IMPLEMENTATION EFFORT

This section discusses the initial steps leading towards successful EMS implementation. Addressed are the early readiness assessment work through the basic organizational activities, such as the staffing for the effort and decisions regarding the parts of the organization to be covered by the initial EMS activities. Figure 4.2 illustrates some of the initial activities, and also provides a look ahead at how these initial activities fit into future implementation steps discussed in Chapter 5.

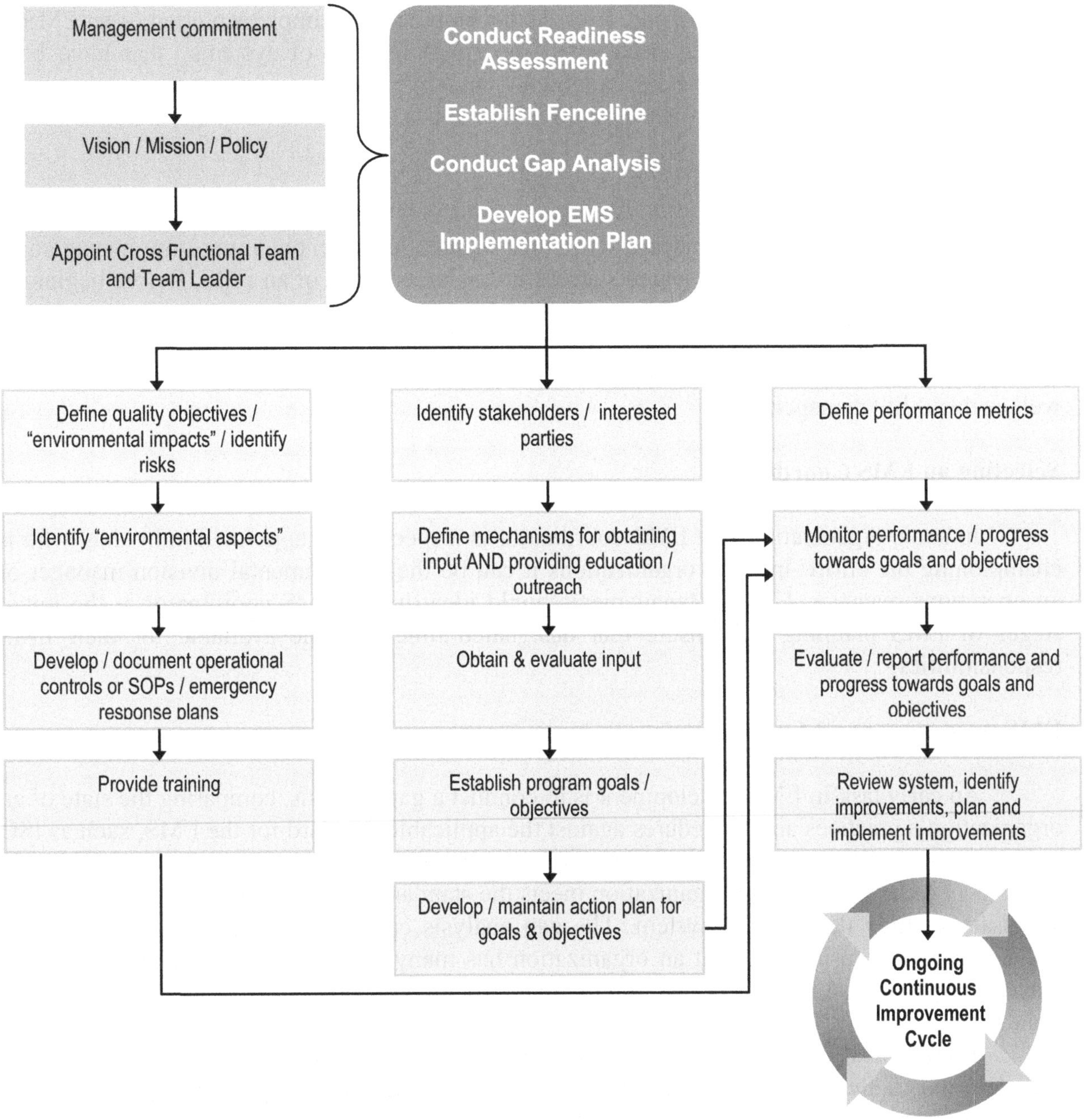

Figure 4.2 EMS implementation process map

Estimating Required Time and Resources

EMS development and implementation does take time and resources, primarily in the form of labor hours. Some organizations use consultants who specialize in EMS implementation to help with the extra resources required during EMS planning and implementation. Other organizations rely on existing staff by reallocating resources, redirecting responsibilities, and/or reprioritizing other projects. Many organizations use a combination of consultants and staff time. When utilizing consultants, it is important to keep staff closely involved so that the system is ultimately owned by and useful for the organization and employees.

Information on resources required for the development and implementation of an EMS in drinking water utilities is limited because of the small number of systems that have been established. The EPA wastewater handbook contains useful information on resource requirements from several wastewater facilities (EPA 2004).

Top Management Commitment

As mentioned above, under Factors for EMS Success, obtaining top management commitment is an important first step to successful implementation of an EMS. Before beginning the implementation effort, the staff promoting EMS will need to educate management on the reasons for EMS implementation and make a convincing case. In a public utility, top management can include elected officials such as city councilors or utility board members, as well as the utility management.

Selecting an EMS Coordinator

In some organizations the EMS coordinator is a specifically appointed individual who is championing the effort. In other organizations it can be the environmental division manager or an operations manager. Utility Management should identify the EMS coordinator at the initial stages of EMS planning and ensure that designated time is made available for their EMS responsibilities.

EMS Assessment and Gap Analysis

An early task in EMS development is to conduct a gap analysis, comparing the state of an organization's practices and procedures against the applicable standard for the EMS, such as ISO 14001. This gap analysis, which can be conducted in house or by a consultant, provides an assessment of how closely the organization meets the standard and identifies the amount of work necessary to fully develop the system. The gap analysis can also be a useful morale booster because it can often indicate that an organization has many of the EMS ingredients already in place; ingredients that may need only fine tuning or documentation to be fully compliant with the standard.

The gap analysis results can be summarized in a table in which the level of conforming to the elements of the standard is indicated by categories such as "fully compliant," "partially compliant," "minimally compliant," or "element does not exist."

EMS Team Members

Most organizations organize an EMS team charged with responsibility for designing and implementing the system. Ideally, this "core team" should comprise six to 10 individuals from all levels in the organization, including line staff. The team should be supported by top management and be given the time and resources to do their work. The core team should also receive the necessary EMS training. Some organizations use a consultant to provide this initial training and initial direction

In addition to providing overall leadership to the EMS effort and being the group visibly working on the EMS project, the core team is responsible for developing EMS documents and procedures and significant aspects, objectives, and targets; providing training materials; and reporting back to respective work sections.

Larger utilities may find it useful to designate additional teams for functional areas such as operations or maintenance. These additional teams will have responsibilities for developing work-area-specific EMS components such as standard operating procedures (SOPs).

Defining the EMS "Fenceline"

Prior to embarking on the planning work, the utility must make a decision about the scope of the proposed EMS—to what parts of the organization will the EMS apply? This decision will influence all subsequent development work. The EMS term used is "fenceline." The fenceline can define a physical boundary (the EMS will include one facility but not another) or an organizational boundary (a single operation in a facility or a department). The fenceline should also make sense for each organization—merely copying another organization's fenceline is not a good idea. Key considerations to consider include the following:

- ***Understanding organizational structure, areas of responsibility, and financial structure.*** The organizational structure may lead to an obvious definition of the fenceline. It is important that there be a clear line of overall responsibility for the EMS. How the organization is funded may also be important—if parts are funded from different sources, that may be a factor in the fenceline decision. Business function could also be used to determine the fenceline—some utilities start with just operation and maintenance business functions and then later branch out to include other business functions like the lab, human resources, and purchasing. There is potentially some overlap in these ways of approaching fenceline. For example, one could think of the lab as a business unit or as a physical entity.
- ***Starting small, then expanding—phased implementation.*** If the organization wants to apply EMS principles to a single aspect before taking it to the whole organization, choosing a small part of the organization—such as the laboratory—to pilot the EMS can make sense.
- ***Determining which parts of the organization have shown interest.*** If the EMS initiative is originating in one part of the organization, this may be a logical initial fenceline.
- ***Deciding whether there are environmental issues that need to be managed better.*** In some organizations there are areas that have ongoing or hard-to-solve environmental issues that merit extra attention.
- ***Focusing on areas that have greater environmental impacts—energy use, hazardous materials use, natural resources consumption.*** Again, if these areas exist, they might be logical choices for the initial fenceline.
- ***Determining whether there is a particularly receptive unit.*** Is there a business unit that, for whatever reason, is most ready for successfully launching a continual improvement culture?
- ***Gathering information from other initiatives.*** For example, if the organization has already completed some QualServe benchmarking, the EMS could be piloted in the part of the organization where the benchmarking indicated the greatest opportunity for improvement.

- ***Considering the size of the utility.*** If the organization is small, there is likely only one treatment facility, so the fenceline might include the entire facility. However, there may be greater resource constraints, so a stepwise approach might make sense. Larger utilities typically have more resources but may need to expand the fenceline in steps because of the bureaucracy or other factors that make implementing change at a large organization difficult.
- ***Gauging rural and urban differences.*** Utilities located in and serving urban areas might have different types of public concerns that could influence how a utility establishes the fenceline.

INTEGRATING EMS WITH OTHER PROGRAMS

There are a number of voluntary programs and tools available for water utilities that are designed to support performance improvement. Hundreds of utilities are participating in voluntary programs such as the Partnership for Safe Water and QualServe and using tools such as the Balanced Scorecard. Utilities should consider building upon and linking any existing environmental or quality programs and tools to the EMS, rather than looking at them as unrelated or even competing activities. Utilities that are participating in these programs can incorporate goals, performance targets, plans and activities into the EMS. For example, Partnership for Safe Water performance targets for drinking water turbidity can be included under EMS objectives and targets.

When embarking on EMS development, utilities might begin by identifying which elements of the plan-do-check-act cycle the other tools and initiatives might already support and incorporate them into EMS elements. For example, utilities might consider if performance goals or targets have already been established, or if voluntary requirements committed to under existing programs should be included under EMS legal and other requirements.

Experience has also shown that utilities embarking on an EMS can also make effective use of key elements of other voluntary programs. For example, an organization could use the best management practices and benchmarking components of QualServe to develop its EMS. While an EMS is intended to go beyond most other voluntary programs, there is no need for an organization to discontinue use of these other programs just because it is implementing an EMS.

Each utility will have its own reasons and drivers for choosing to implement an EMS or to adopt a new voluntary program or tool. There is no one set of programs or order for implementation that will fit for all water utilities. Rather, utility managers must determine what is important for their organization and which tools and programs would be support desired performance improvements. Utilities have found that performance improvements with an EMS are only enhanced when combined with other tools and initiatives. There are a number of existing programs and guidance documents available, as well as websites for contacting peers with experience.

Following are brief descriptions of just a few voluntary programs in which a number of utilities are already participating, and how they can be integrated with EMS. More information for water utilities on integrating EMS with strategic planning and voluntary initiatives, including utility case studies, is provided in existing guidance documents cited in the resources at the end of this chapter. These resources also list a variety of EMS guidance documents, tools and voluntary programs designed to improve performance of relevance to water utilities.

QualServe

QualServe is a voluntary improvement program designed specifically for water and wastewater utilities. QualServe includes self-assessment, peer-review, and benchmarking activities to help utilities compare their practices and performance with other utilities and industry best practices. QualServe is an example of a voluntary commitment that would be identified under "other requirements" of an EMS.

QualServe provides an approach for utilities to perform a high-level evaluation of all of utility operations. Utility managers can employ QualServe to prepare a baseline from where it is starting, which can be utilized in the process of setting strategic direction and policy, as well as setting organizational goals and objectives. In this fashion, QualServe can support the planning phase of developing an environmental management system.

QualServe best practices can be documented and implemented as part of EMS operational control. And, although not specifically designed to support monitoring and measuring, QualServe can support the checking component of a management system framework by using the evaluative tools provided by the program as one way of assessing current practices.

AwwaRF, EPA, and the QualServe Benchmarking Clearinghouse collaborated on a research project, published in 2004, which provided guidance in the selection of appropriate performance indicators for water utilities (Crotty 2004). Twenty performance indicators, grouped by QualServe's five business process categories, were recommended for implementation in that research project. These performance indicators can be incorporated into EMS targets.

Partnership for Safe Water

The Partnership for Safe Water is a voluntary performance program that incorporates benchmarking through data collection. The Partnership's program includes a commitment to continued compliance with the existing Surface Water Treatment Rule, baseline data collection and analysis of existing turbidity data, and continuing annual submittal and analysis of turbidity data. Utility managers who want to focus on drinking water turbidity improvements can implement Partnership for Safe Water by adopting turbidity performance targets, collecting turbidity data to provide a benchmark of utility performance, and continuing an annual cycle of making improvements and collecting turbidity data. How a utility increases turbidity performance through adjustment of policies and practices is up to the individual utility—Partnership for Safe Water does not provide best practices in this regard. The targets provided by the Partnership for Safe Water can be directly incorporated into EMS objectives and targets. The procedures for monitoring and measuring water quality that must be established for PSW can also be incorporated under EMS operational control and measuring and monitoring procedures.

Strategic Business Planning and Long Range Financial Planning

Water utilities that engage in strategic business planning efforts will find that an EMS can be a mechanism for achieving strategic and financial goals relevant to environmental performance. As well, an EMS can provide a process for taking high-level organizational goals from the strategic plan and deriving more specific environmental objectives and targets. An EMS also supports implementing environmental programs and operational controls to achieve those environmental objectives.

National Biosolids Partnership EMS

The NBP has developed an EMS for biosolids, designed for use by wastewater treatment or combined utilities that create and manage biosolids. While loosely based on ISO 14001, the EMS for biosolids focuses specifically on those environmental impacts related to biosolids management activities. Combined utilities that are interested in or are already implementing an EMS for biosolids can expand the fenceline to cover other wastewater treatment and water treatment operations. The EMS procedures developed under an EMS for biosolids can be directly applied to other parts of the organization and will meet most ISO 14001 requirements.

An organization that is interested in starting with an ISO 14001 EMS can similarly take on the extra NBP requirements for biosolids management activities, such as the enhanced public participation requirements. Procedures from the ISO 14001 EMS can be directly applied to biosolids activities and will meet most NBP requirements. Organizations that take this path will find that just a few additions are needed to meet NBP EMS program requirements.

SIMPLIFYING THE APPROACH: CONSIDERING A TOOLKIT APPROACH TO EMS

Several options exist to simplify the EMS development process. One (used by the NBP to assist wastewater utilities) is to create guidance documents, training modules and a network or participating agencies to share their work and lessons learned. Another is to create a model EMS manual with menu selections for key areas such as environmental aspects and impacts. This approach would improve consistency with EMS standards and reduce the effort for interpreting guidance documents to develop a functional EMS. A third option is to approach EMS development by components rather than as a complete system. This might reduce barriers to implementation for a small utility that is interested in an EMS but intimidated by the thought of developing a comprehensive EMS. These components could be later linked together to form the system. Some examples are listed below:

- Identifying priorities for improvement, developing effective performance indicators, and setting targets for improvement
- Establishing operational control of key aspects or activities (e.g., by determining what needs to be managed effectively, and developing procedures to manage them)
- Identifying risk scenarios and reviewing the effectiveness of existing prevention and response procedures and filling any gaps.

These options offer agencies some alternatives to the cost, time and resources required to develop an EMS from scratch. Agencies considering developing an EMS may want to consider and explore the potential benefits and efficiencies these options may provide.

RESOURCES FOR MORE INFORMATION

Achieving Environmental Excellence: An EMS Handbook for Wastewater Utilities. See www.peercenter.net/sector/wastewater. Prepared by EPA and the Global Environment and Technology Foundation (GETF). August, 2004.

Balanced Scorecard. See http://www.balancedscorecard.org/. See also *Translating Strategy into Action: The Balanced Scorecard,* by Robert S. Kaplan and David P. Norton, Harvard Business School Press, 1996.

EMS. See Multi-State Working Group. *The External Value Environmental Management System Voluntary Guidance (DRAFT).* August 29, 2003. Prepared by the MSWG EVEMS Subcommittee. (February 3, 2004; http://www.mswg.org/EVEMSdraft8.29.03.pdf).,

EMS Guide for Small and Medium Sized Organizations. Philip J. Stapleton, Margaret A. Glover, and S. Petie Davis. *Environmental Management Systems: An Implementation Guide for Small and Medium-Sized Organizations (Second Edition).* Ann Arbor, MI: NSF International, 2001.

EMS and Local Government. See *Environmental Management Systems: A Guidebook for Improving Energy and Environmental Performance in Local Government*, by Five Winds International/PA DEP, 2004. See also *An Environmental Management System Troubleshooters' Guide for Local Governments,* by Global Environment & Technology Foundation, 2002.

EPA Performance Track. See http://www.epa.gov/performancetrack/. Example of an Annual Performance Report is available at https://yosemite.epa.gov/opei/ptapr1_0.nsf/392fb8cedd796f8a85256b8e007da7bb/6c122a09c9ddb0de85256bf10079366f?OpenDocument.

Global Reporting Initiative. See http://www.globalreporting.org/.

Integrating EMS. *Moving Toward Comprehensive Utility Management Systems—Report of the Environmental Management Systems (EMS) for Public Utilities Integration Project* (2002) and *Continual Improvement in Utility Management: A Framework for Integration* (2004) published by NACWA, EPA, and WEF.

ISO 14001. Lexington Group/USAID Office of Energy, Environment and Technology. *Best Practices Guide: Application of ISO 14000 Environmental Management Systems (EMS) for Municipalities*. Undated.

National Biosolids Partnership. See http://www.biosolids.org.

Partnership for Safe Water. See http://www.awwa.org/science/partnership/.

Peer Center. See http://www.peercenter.net.

Performance Indicators. Crotty, Pat. *Selection and Definition of Performance Indicators for Water and Wastewater Utilities.* AwwaRF and AWWA. 2004.

Performance Measurement. See *Developing and Implementing a Performance Measurement System* (vol. 1), by Water Environment Research Foundation, 2000.

QualServe. See http://www.awwa.org/science/qualserve/.

CHAPTER 5
SPECIFIC STRATEGIES FOR PLANNING, DEVELOPING, AND IMPLEMENTING AN EMS

This chapter contains specific guidance on the EMS elements and includes considerations for drinking water utilities in developing the element. A list of resources for additional information is included at the end of the chapter. When available, examples of EMS elements, tools and templates have been provided in Appendix A. These examples are from actual ISO 14001 EMS manuals and procedures provided by the Charleston, South Carolina, Commissioners of Public Works (CPW); Chesterfield County, VA, Elizabethtown Water Company (American Water); the City of Gastonia, NC; and the City of San Antonio, TX.

ENVIRONMENTAL POLICY

Outline of Element, ISO Requirements, Purpose and Relationship to Other Elements

The environmental policy lays out the utility's commitment to protecting the environment and public health. It drives not only the EMS but how the utility operates and how it makes decisions regarding the environment and public health.

The ISO standard 14001 requires that the environmental policy contain at a minimum three basic commitments:

- Continual improvement
- Prevention of pollution
- Compliance with legal requirements (relevant environmental laws and regulations)

Beyond this minimum, the standard includes some additional requirements: that the policy be made available to all employees and the public, that it provide the basis for setting objectives and targets, and that it be appropriate for the organization.

The environmental policy element helps drive the EMS because the rest of the EMS will be built upon the principles articulated in a utility's commitment to the environment and public health. The importance of the policy is realized in the ISO expectation that all staff from all levels of the utility will know the policy, understand what it means to the utility, and be able to explain how they promote it in their day-to-day functions.

Considerations for Drinking Water Utilities

In addition to the mission of cost-effective production of a product or delivery of a service that is shared with other industrial sectors that have implemented environmental management systems, drinking water utilities also have the additional primary mission of *protecting* the public health. The key difference here is that there is an inherent environmental aspect to the mission of the drinking water utility that does not exist in most other sectors. Therefore, this mission to protect public health will need to be included in the environmental policy for drinking water utilities. While this may differ from other industries' environmental policies, it should not be difficult for drinking water utilities to incorporate, as they have been fulfilling this mission for decades. As long as a utility accurately reflects its commitment to public health in its environmental policy, it should work as an effective base from which to build its EMS.

Another consideration for drinking water utilities is that their governing and management structure may differ significantly from other industry sectors, especially for publicly owned and operated water systems. There may be jurisdictional limits, policy-setting limits, and other boundaries in place that do not exist with private-sector companies. This is an important consideration because it will affect those who need to be involved in the environmental policy development and approval, as well as how long it may take to obtain approval. As the policy is the foundation to the development of the EMS, it is important that these considerations be taken into account in the initial stages of EMS planning, and efforts taken to minimize their impact on the overall EMS development.

Some upfront efforts that may make the policy development and approval easier include identifying who will need to approve the policy and mechanism for obtaining approval, and to note how this will affect the schedule. For example, in certain jurisdictions, policies can be approved only by the governing body such as the city council, or board of county commissioners. In these instances it will be important to take into account how hands on and actively involved this governing body is to understand how much upfront information and questioning they will have before granting approval to the environmental policy. For scheduling purposes, it will also be important to understand how items make it onto the agenda for approval. This process can take several weeks or months as items are required to be placed on planning meeting or advisory board agendas before they can be placed on the official governing body agenda for approval.

How to Get Started

The following are suggestions about how to get started with developing and approving the environmental policy.

Determine who will be involved with the policy development and approval. This will break down into three categories:

1. Staff who are directly involved. This will be the utility's EMS team. These people will be responsible for drafting and obtaining approval of the policy.
2. All other staff. Because the policy will impact everyone at the utility it is important that input and buy-in on the policy is received from the staff. The best way to do this is upfront before the policy is sent for approval. Review the operational areas that will be involved in the EMS and bring in appropriate personnel from these areas for input. This may also be the time that staff education and preparation for EMS development and implementation begins; doing so now may make things easier later on. Types of input and questions that may be asked of these personnel are the following: (a) What do you think is the primary mission of the organization? (b) Do you believe that your job supports this mission? (c) Do you think that anything is missing from the mission of the organization? After this input is gathered and the policy has been developed, allow these personnel to review and comment on the policy to ensure that it accurately reflects the mission and operation of the utility.

3. Individuals or groups responsible for approving the policy. This will be the individual members of the governing body and the governing body as a whole that has the authority to approve policy for the organization.

Think about the environmental policy development and approval timeline. Approval may take a lot of time depending upon management operations. After identifying the approval parties and the individuals that should provide input, develop a schedule to toward obtaining policy approval. This schedule should include time for (1) gathering input, (2) drafting policy, (3) reviewing policy, (4) and approving policy.

Review examples from similar utilities before drafting the utility's policy. While your content may be different to reflect your mission and operations, examples from other utilities may provide useful ideas for content and structure.

Implementation Recommendations, Tips, and Suggestions

The policy should describe the utility's core values, what the organization "stands for." The utility's actions, goals, and performance should be easy to associate with the principles articulated in the policy. Similarly, there should be no contradictions between actions (or absence of actions) and policies. However, it is also important to recognize that some balancing of values is inevitable. For example, "providing affordable service to customers" may need to be balanced with "providing sector-leading service to customers."

Simplicity is often the best guideline for developing an environmental policy. A goal of one page, with a maximum of two, will promote the use of clear and concise language that can be understood and remembered by staff. It is often effective to freely collect input and then condense the material later.

Many organizations print wallet-sized cards for employees that summarize the environmental policy. This helps signal that the policy is intended to be used to guide day-to-day decisions, and allows employees to refer to the card for guidance and during an audit.

Examples, Tools, and Templates

Appendix A.1 contains an example environmental policy statement from the Charleston CPW that includes the CPW's commitment to improvement of the environment and to the establishment and maintenance of an EMS that is compliant with ISO 14001. The policy statement also includes the utility's mission, vision, strategic business plan and core values of the CPW. This is a useful example because it illustrates not only how a one-page policy can be compliant with the ISO 14001 requirements, but also how it supports the mission and vision of the utility.

ENVIRONMENTAL ASPECTS AND IMPACTS

Outline of Element, ISO Requirements, Purpose and Relationship to Other Elements

In this part of the planning process for system development, an organization undertakes a review of its activities, products, or services and determines which of those contributes to an environmental impact. Additionally, ISO 14001 requires an organization to prioritize these environmental impacts, and define those that are significant. Further, the standard requires that these significant impacts are considered when the organization sets its environmental objectives, and also requires that the information on environmental aspects and impacts is kept up-to-date—i.e., is periodically reviewed and revised if necessary.

This element is intended to promote an exercise through which the organization identifies its environmental impacts, and as a result goes through an educational process. Frequently this exercise will uncover environmental impacts that have previously not been understood or acknowledged.

As a fundamental element in the EMS, it has strong links to other elements: identification of significant aspects influences the development of objectives and targets, environmental management programs, and monitoring and measurement programs.

Considerations for Drinking Water Utilities

As discussed previously, drinking water utilities differ from a typical manufacturing facility in many ways. A drinking water facility can be viewed as having one major "product"—finished water—and several minor by-products, including residuals.

The "activities" of a drinking water facility include many of those that have been previously documented in the literature for wastewater treatment plants, including activities associated with office, maintenance, transportation, laboratory, pipe networks, etc.

The "fenceline" established for the EMS (Chapter 4) will determine the scope of the aspects and impacts inventory. For example, if a utility chooses to limit the fenceline of the EMS to the actual footprint of the treatment facility, then impacts resulting from source water withdrawal, or distribution system construction and operation need not be included. However, decisions regarding how the fenceline is defined need to consider commitments implied or stated by the Environmental Policy.

How to Get Started

The development of an inventory of aspects and impacts generally follows the following sequence:

- Identification and chartering of team charged with development of the aspects and impacts inventory
- Determination of "fenceline," if not already defined
- Analyzing operations and activities within the fenceline—construct process flow diagrams
- Developing an inventory (list) of aspects and impacts
- Determining the significance criteria and apply the criteria to the list of aspects and impacts
- Developing the inventory of aspects and impacts selected as significant aspects
- Documenting the procedure that was used for this exercise
- Repeating the process at regular intervals, identifying any changes and their effect

Implementation Recommendations, Tips, Suggestions

Jargon can be a barrier to effective communication, and one of the hardest terms for staff in a utility to understand is "aspect." Even people who have worked with EMS principles for years have trouble with this term. Since an environmental aspect is really the "cause" of an impact, consider replacing "aspect" with "cause" when speaking or writing about aspects and impacts. The inventory will then consist of a list of activities, with corresponding environmental impacts, and the associated cause of the impact. Completing the sentence, "An (environmental impact) is caused by (cause/aspect) associated with (activity)" may help clarify this association of terms. For example, air pollution (an impact) is caused by burning fuel (an aspect) associated with vehicle use (an activity).

There is no need to start from scratch—use the experience of other utilities that have been through this exercise. As indicated below, included in this guidance are several examples of completed aspects and impacts inventories, and significance ratings systems. These examples can provide a springboard for moving forward with this work.

Aspects and Impacts Identification

This exercise, which depends on the utility's fenceline decision (see Chapter 4) is best conducted by a staff team from the applicable work area—for example, maintenance staff should conduct the inventory for the maintenance work section's activities (with necessary direction/coaching from the EMS coordinator). Typical steps in this exercise include:

- List activities (e.g., parts cleaning)
- Add associated aspects (e.g., waste solvent disposal)
- Add associated potential environmental impacts (e.g., potential water or land pollution)

Inventories from the work areas are then combined into a master list for the facility (or fenceline). At this point duplicate entries can be deleted—for example, all work areas will probably report aspects/impacts from vehicle use or office activities.

Aspects and Impacts Significance Analysis

Environmental aspects need to be prioritized to determine the utility's list of "significant" aspects. Ranking methods can be based on environmental or business issues, or both. Environmental issues ranking methods could include scale of impacts, severity of impact, probability of occurrence (sense of expected value), duration of occurrence, (how long will the adverse environmental consequence persist in the environment), etc. Business issues ranking methods could include cost of changing impact, legal or technical (ease of changing impact), community (concerns of interested parties), and effect on public image.

An example list of aspect evaluation criteria from a drinking water utility may include:

- Normal, abnormal, or emergency
- Legal obligation or company policy
- Impact likelihood

- Impact severity
- Prosecution/complaints
- Easy win (company has direct control or not)?
- Impact total is equal to product of likelihood, severity, and prosecution/complaints

Note that in many environmental management systems, the significance analysis is set up so that when a significant aspect has control measures developed and implemented (such as an operational control procedure), then the aspect is no longer significant, and drops off the inventory. In this case, the utility would usually add another aspect to the significant aspects inventory.

Examples, Tools, and Templates

Appendix A.2 contains example aspects identification templates, as well as an example environmental aspects ranking and grading templates from the Charleston CPW, Chesterfield County, VA and the Elizabethtown Water Company. The aspects identification templates illustrate how different templates can be used, and still prompt the utility to meet all of the ISO 14001 requirements for aspects and impacts. These templates also demonstrates how aspects and impacts are closely linked to other ISO 14001 elements by including control method(s) as well as legal and other requirements. The environmental aspects ranking and grading templates are good examples of how to meet the ISO 14001 requirement that the environmental aspects be prioritized. These three examples also show the wide range of the types of aspects different water utilities encounter.

LEGAL AND OTHER REQUIREMENTS

Outline of Element, ISO Requirements, Purpose and Relationship to Other Elements

The purpose of this element is to ensure that the organization maintains compliance with legal and other environmental requirements. While this is something that most drinking water utilities manage very well, having it as an element of a functional EMS ensures that the organization stays abreast of existing and upcoming requirements even during times of personnel changes or unusual operations.

This is the only element in ISO 14001 that specifically requires the organization to establish and maintain a procedure. In this case, the procedure is required to identify the legal requirements that apply to the organization's activities, products, and services. The procedure must be implemented, and the list of legal requirements must be maintained. The "legal and other requirements" element is closely linked with the documentation, document control, and recordkeeping element, which covers time periods over which data and records must be kept. Therefore, the identified legal requirements should be referenced the document control and recordkeeping element, as well.

Considerations for Drinking Water Utilities

One unique consideration for this element is that public health requirements will be included as part of the environmental legal and other requirements. This means that drinking water utilities may have more requirements to track than other industries do.

How to Get Started

An inventory of legal requirements is generally developed in the following sequence:

- Identify the personnel responsible for tracking legal and other requirements
- Develop a list of legal and other requirements
- Document how legal and other requirements are tracked and records are kept in a procedure for the element
- Review any upcoming changes to legal and other requirements and determine if anything coming up may require operational changes that should be included in objectives and targets

Also included in the legal and other requirements elements for a publicly owned utility will be the applicable municipal codes, policies and rules.

Implementation Recommendations, Tips, and Suggestions

It is likely that the utility already has a process in place to track legal and other requirements. It is not necessary to rewrite this process if it is working well, but the process should be clearly documented to reflect what is actually done. Included in Appendix A.3 is an example of one utility's approach to this element.

Examples, Tools, and Templates

Appendix A.3 contains an example EMS procedure for legal and other requirements from San Antonio, TX. This is a useful example because it illustrates a straightforward procedure for meeting the ISO 14001 requirements for this element. This procedure also shows the other important parts of an EMS procedure such as scope and purpose, recording of the revision date and revision tracking and procedure approval.

OBJECTIVES, TARGETS, AND PERFORMANCE-BASED MANAGEMENT

Outline of Element, ISO Requirements, and Purpose and Relationship to Other Elements

As part of the EMS planning, an organization must establish and maintain documented objectives and targets. Objectives and targets define how the organization will pursue continual improvement under its EMS, including the specific environmental performance improvements the organization is working towards. Objectives and targets support the achievement of legal and other requirements and must be consistent with the environmental policy, including the commitment of pollution prevention.

In setting objectives and targets, the organization must consider:

- Legal and other requirements
- Technological options
- Significant environmental aspects
- Financial, operational and business requirements
- Views of interested parties

Environmental programs, training, and operational controls should all be aligned to support achievement of objectives and targets. Performance reports and communications demonstrate progress towards objectives and target to internal and external audiences.

During the management review, goals and objectives are examined and updated on a periodic basis. This management review and update may launch corrective actions necessary for performance improvement.

Objectives are broadly stated goals for environmental improvements; targets are detailed, measurable performance requirements necessary to achieve environmental objectives.

How to Get Started

Moving from Significant Aspects and Environmental Impacts to Objectives and Targets

Utilities should manage all significant environmental impacts that are identified during the aspects and impacts analysis, but set objectives and targets only for environmental issues that the utility wants to improve. Objectives and targets should be focused on what the utility wants to achieve. Utilities may find that the highly significant aspects already have operational controls and best practices in place, in which case they just need documentation. In addition to considering those things that are mandatory, such as legal and other requirements and significant environmental aspects and impacts, utilities also consider additional factors as a basis for choosing where to set objectives and targets, such as:

- Easily accomplished actions for early success
- What is realistically feasible
- High expected value
- Politically sensitive areas
- Cost and potential return on investment
- Employee input
- Strategic planning goals

Implementation Recommendations, Tips, and Suggestions

Employees need to understand goals and how they impact them. Input from interested parties can be obtained by the EMS team members who participate on different committees (watershed groups, etc.). If at all possible, consider linking an incentive program to encourage and reward employees focused on goals.

Communications and Objectives and Targets

As part of EMS communications, utilities will want to make sure that the objectives and targets are responsive to the community and that they help generate support for the EMS. When launching an EMS, utilities need to have a candid discussion with staff and stakeholders about goals for improvement. If there is intense external scrutiny of the utility, seeking external input for goals may need to be managed very carefully. Additional considerations include:

- Carefully address how the objectives and benefits of the EMS are communicated if it is being implemented within a period of perceived problems.
- Understand that the utility's EMS team has an initial learning curve and will not be experts immediately; do not expect them to be able to explain all details of the EMS to staff or outside audiences without some accumulated experience
- Gain some experience with the EMS before trying to set comprehensive goals. The experience with a few issues will help inform your future efforts and let you apply learned lessons.

Measurable Targets

Performance targets should be selected with care to ensure that they serve their intended purpose of measuring progress toward a goal. Utilities should consider using targets that meet "SMART" criteria:

- Specific
- Measurable
- Achievable
- Relevant
- Timely

Examples of SMART targets might be the following:

- Reduce electrical energy use by 5% from 2002 baseline by 2008
- Maintain turbidity levels below of 0.8NTU by 2006
- Achieve and maintain line loss efficiency at 8% by 2010.

Measuring improvements in performance requires documenting changes over time, meaning that measurement of baseline conditions is required. If baseline data do not already exist, utilities can set as a target the collection of baseline data so performance improvement can be measured in the future.

Operational controls (for example SOPs for measuring and monitoring) should be used to ensure that data are collected correctly. Many utilities already have these types of procedures in place, in which case they simply need to be documented as part of the EMS.

Examples, Tools, and Templates

Examples of Areas for Establishing Objectives

Note that these are descriptions of *general* issues, activities, or impacts for which an objective may be established:

- Chemical use (reduction, shift to less risky alternatives)
- Water conservation
- Residuals management
- Source water protection
- Chemical storage, handling, transport, delivery
- Water leakage (in supply or distribution system)
- Fleet management /fuel consumption/emissions
- Security—chlorine, source water protection
- Security—water intake
- Back-up power—fuel storage, permitting, emissions
- Materials management—pollution prevention/waste minimization/recycling
- Energy consumption
- Product quality—taste/odor/contaminants
- In-stream impacts—withdrawal/discharge impacts on water quality, fisheries

Examples of Areas for Establishing Targets

Note that the following examples are not *targets* themselves, but rather are examples of specific issues for which a target might be established:

- Ratepayer perception of utility's stewardship of the environment
- Turbidity levels
- Chlorine residuals
- Recycling or beneficial reuse of materials
- Energy use or supply source ("green" energy)
- Emissions to air, land, water
- Backwash water management
- Water efficiency and conservation
- Chemical use
- Chemical spills frequency and severity
- Ratios of preventive vs. corrective actions
- Incidents requiring "boiled water" notification to users
- Water line breaks
- Work orders completed

Different Types of Measures to Consider

All of the above examples can be appropriate for EMS objectives and targets. In fact, it can be very helpful to measure or identify indirect measures, for example tracking training that is completed. Other considerations include:

- Internal vs. external (internal improvements would be assessments of a utility's own operation, while external measurements might be a comparison to other utilities for benchmarking)
- Direct vs. indirect (a direct measure of quality operations may be ISO 14001 certification, while an indirect measure may be number of operators who have received more than 20 hours of training per year)
- Output vs. outcomes (activities and work products are outputs, which may be easier to measure, while results are outcomes)

Appendix A.4 contains example objectives and targets for a number of aspects identified by Chesterfield County, VA. These are useful examples because they demonstrate how the objectives and targets tie back to the environmental aspects, as well as how each objective and target needs a management program (see next section).

ENVIRONMENTAL MANAGEMENT PROGRAMS

Outline of Element, ISO Requirements, and Purpose and Relationship to Other Elements

This element in the EMS results from the work conducted under the Environmental Aspects, and Objectives and Targets elements, and is also related to the Management Review element.

The ISO 14001 standard requires the organization to develop Environmental Management Programs (EMPs) for achieving objectives and targets. The programs must include details such as responsible parties, timelines, resources, etc., and can be viewed as "action plans" or "work plans." The EMPs are a description of how the facility will work towards meeting objectives and targets.

Considerations for Drinking Water Utilities

Prior to developing Objectives and Targets, and associated EMPs, a utility will need to consider whether these programs should relate to product quality (high public interest) or other environmental targets (possibly less public interest) or both. In addition, the utility will need to consider how the objectives and targets will be communicated with the public. This could be combined with existing reporting mechanisms.

How to Get Started

- Develop a template for EMPs, or use another utility's as a model
- For each objective and target, designate a "manager" who takes overall responsibility for meeting the objectives and targets. Often there is a logical choice in an

organization for identifying these managers. For example, if the objective/target relates to facility operational use of energy, the operations manager would be a good choice. For utilities with multiple facilities—a person from each facility can be designated as "manager."

- May need upper management commitment early if EMP will require significant staff or financial resources
- Anyone assigned responsibility in an EMP needs to be involved in program development

Examples, Tools, and Templates

Appendix A.5 contains three example EMPs from the Elizabethtown Water Company. These examples illustrate a good format for developing EMPs and depict the type of information that should be included in an EMP (e.g., environmental objective and related targets, the party(ies) responsible for completing the EMP, the related environmental aspects and impacts and needed resources).

STRUCTURE AND RESPONSIBILITY

Outline of Element, ISO Requirements, Purpose and Relationship to Other Elements

The Structure and Responsibility element deals with the organizational structure and the lines of responsibility, as they relate to the environmental management system. ISO 14001 requires that the roles, responsibilities, and lines of authority be defined by the organization, and then documented and communicated to the employees. Also included in this element is a requirement that management provide the resources necessary for implementation and maintenance of the EMS, such as employees with any necessary specialized skills, technology, equipment, and financial resources.

This element requires management to appoint staff to the key roles in the EMS (see Chapter 4) and designating the responsibility, along with the appropriate authority, for ensuring that the EMS is developed and implemented according to the requirements of the standard, and also for reporting on EMS performance to management.

How to Get Started

When the planning phase of the EMS development begins, consider the current organizational structure of the facility for which the EMS will be developed. The size of the facility plays a role—a small facility may not need very much rigidity in its structure, whereas a large facility will need more detailed structure to clearly assign responsibility, and to prevent important tasks and responsibilities from falling through the cracks. The key organizational roles have been previously discussed in Chapter 4; this element requires that these roles be documented.

Examples, Tools, and Templates

Appendix A.6 contains an example Structure and Responsibility EMS procedure for the Charleston CPW that defines how the structure and responsibilities associated with Charleston CPW's EMS program are established. This example illustrates how a simple procedure can apply to all of a utility's facilities within a single utility-wide EMS,

TRAINING, AWARENESS, AND COMPETENCE

Outline of Element, ISO Requirements, Purpose and Relationship to Other Elements

The ISO 14001 standard for this element requires the organization to identify the training needs of its employees, and requires appropriate training for any employee whose work may create a significant environmental impact.

In addition, the standard includes a list of specific items for which all employees need to have awareness. These items include knowledge and understanding of the organization's environmental policy; the significant environmental impacts of the employee's work activities; the employee's roles and responsibilities for implementing and maintaining the EMS; and the potential environmental consequences of not following EMS procedures.

How to Get Started

This element of the EMS has three components—training, awareness, and competence. A good place to start is with the awareness component. This activity can begin in the early planning stages of the EMS. A technique successfully used by many organizations is to create an EMS newsletter. Early issues can be limited to basic information; later issues can describe the progress being made to develop the EMS. The newsletter can be distributed through e-mail, posted on bulletin boards, or printed as a flyer.

When the training phase begins, the EMS training can be integrated with other existing training programs, safety for example. EMS training can be divided into two components—the basic training that every employee should receive (policy, overall awareness of significant aspects, objectives and targets, etc.) and the more specific training dealing with the environmental aspects and impacts of the employee's particular work area. An on-going training program should also be developed to keep employees current on the EMS performance and any changes to the system (revised targets, etc). This can be accomplished through an annual refresher training session.

Training does not have to be classroom-style lectures. Other techniques can be successful—such as use of posters; quick, short e-mails with a nugget of information; games with EMS questions and prizes; tailgate sessions; etc.

New employees should receive basic EMS information as part of their orientation, and employees moving from one job area to should be considered for site-specific training.

Examples, Tools, and Templates

Appendix A.7 contains an example training EMS procedure for the Charleston CPW that defines training requirements. This illustrates not only the training needs and requirements, but also how training results in employee competence and how training records must be tracked.

COMMUNICATIONS: INTERNAL AND EXTERNAL

Outline of Element, ISO Requirements, Purpose and Relationship to Other EMS Elements

The communications element has special significance in EMS implementation. Organizations must establish procedures and programs for internal and external communications about the EMS, as well as the organization's environmental policy and environmental performance. Communications ensure that internal audiences such as management, employees, and contractors understand the EMS and their roles and responsibilities with respect to managing environmental aspects and impacts. External communications ensure that mechanisms exist for receiving and providing information about the EMS and environmental performance to affected and interested parties, such as rate payers and the media.

For internal communications, organizations must establish and maintain procedures for communication between the various levels and functions of the organization. The communications element also provides the means by which information relevant to many of the EMS elements is communicated and received.

External communications should provide a mechanism to:

- Make the environmental policy available to the public
- Communicate with external interested parties regarding environmental aspects
- Solicit the views of interested parties for consideration when establishing goals and objectives

For external communications, organizations must:

- Establish and maintain procedures for receiving, documenting, and responding to relevant communication from external interested parties regarding its environmental management system
- Consider methods for external communication on its significant environmental aspects and record its decision

Considerations for Drinking Water Utilities

Internal Communications

It is not uncommon for water utilities to have contractors responsible for certain operations and activities. When establishing internal communication mechanisms, remember to include these contractors and consider their relationship to the organization's environmental aspects as well as their interaction with the public and other external interested parties.

To operate around the clock, water utilities must have multiple shifts of employees. Assure that personnel on all shifts receive information and have the opportunity to participate.

External Communications

An EMS can help an organization address a variety of communication needs. Through the cycle of checking and review built into an EMS, an organization gathers valuable information about environmental performance and management activities that can be communicated to a variety of external audiences. An EMS also establishes procedures for receiving information from external parties which can help an organization identify and understand community values and expectations.

Some of the needs specific to water utilities that an EMS could help address includes the following:

- Relations with local community and environmental organizations who are concerned about water quality and source water protection, such as a local river watch group or watershed protection council. Through an EMS an organization can communicate how it is protecting the watershed and local water quality
- Communications with neighbors and the communities served about projects, operational changes, justification for rates, and commitments for improvement
- Feedback from neighbors and the public about local community values and interests
- Communications with elected officials about service reliability, why costs and rates are what they are, and what the organization is doing to protect the environment
- Communications with businesses, trade organizations, rate review boards
- Communications with regulatory agencies about what the organizations is going to be doing and what it has achieved
- Information to the media that the utility is being a good steward of the environment (commitments and improvements)
- Public education programs to enhance public trust and credibility with customers and stakeholders relative to product quality and security

How to Get Started

It is probably easiest to consider external and internal communications separately when identifying who are the audiences, what needs to be communicated, and what are appropriate communications methods and formats. However, when implementing a communications program, organizations may find overlap in these items that allows for using some communications methods for multiple audiences. For example, a web site may provide information to the public as well as the media and local community and environmental organizations.

Common steps for establishing EMS communications procedures and programs include:

- Identifying internal and external audiences that need to be communicated with; consider who is currently communicated with and who else might have a potential interest in the organization and its environmental management.

- Determining what information needs to be communicated for each audience; consider (a) what information is currently made available, (b) what information can be provided by the EMS (e.g., progress towards environmental improvement goals and objectives), and (c) what information is important to receive from external audiences.
- Determining the methods and formats by which to best communicate; consider what methods and formats already exist and how they can be incorporated into the EMS.
- Creating (or updating) communications programs and procedures that conform to EMS requirements.
- Maintaining records demonstrating the programs and procedures have been implemented (e.g., attendance sheets from public meetings, complaint response logs).

Internal and External Audiences

Some typical audiences for internal communications are senior executives, management and employees, employee unions, and contractors and vendors.

Some typical audiences for external communications include the following:

- Rate payers and other citizens of communities served by water utility
- Neighbors of the utility's facilities
- Community and environmental organizations
- Advisory boards
- Elected officials
- Businesses and trade organizations
- Rate review boards
- Regulators (federal, state, and local)
- Emergency response organizations
- Media and trade press

Information and Messages

Examples of information communicated across internal audiences include the following:

- General EMS awareness, including the environmental policy statement
- Environmental roles and responsibilities
- Legal requirements
- Practices and procedures for operational control
- Emergency response plans and procedures
- Relevant document control procedures

Examples of information communicated to external audiences include the following:

- Environmental policy and commitments
- Environmental management programs
- Progress towards objectives and targets
- Environmental improvement successes

- Planned improvements and benefits
- Opportunities for involvement

Examples of information gathered from external audiences include the following:

- Input on objectives and targets
- Input on environmental aspects and impacts (optional)
- Concerns and suggestions

Methods and Formats

Examples of internal communications methods and formats:

- Utility's intranet site
- Newsletters
- Bulletin board postings
- Brown bag lunches
- Training programs

Examples of external communications methods and formats:

- Bill inserts
- Newsletters
- Water quality reports
- Public meetings
- Advisory group/board meetings
- Internet website
- Facility tours and educational programs

Implementation Recommendations, Tips, and Suggestions

Build on What Already Exists

It is not always necessary to create new communications programs and mechanisms. Utilities can build on the communication mechanisms already in use and incorporate them into the EMS. For example, consider incorporating EMS information into the utility's annual Consumer Confidence Report rather than creating an additional publication.

Remember to Cover the Entire EMS Fenceline

Reviewing operations within the entire EMS fenceline will help a utility identify all internal audiences. It can also be a prompt to confirm where and which internal audiences interact with external audiences such as the public. For example, contracted operations may require contractors to be in a position of communicating with external audiences on behalf of the utility. It is important that operational contractors understand their communication roles and responsibilities and have access to relevant information.

Reactive versus Proactive Strategies

Consider utility needs for proactive engagement of stakeholders and interested parties consistent with local circumstances. Implement ongoing two-way communications with interested parties, which include robust mechanisms for receiving information about external party concerns and interests, as well as proactively providing information and reporting to the public. With a proactive communications strategy, a utility reaches out to its external audiences rather than waiting for audiences to call and complain (reactive strategies).

Why implement proactive strategies? Proactive communication strategies can help organizations meet the needs of, build more positive and collaborative relationships with, and build credibility with the public, stakeholders, and other external audiences.

Examples, Tools, and Templates

Appendix A.8 contains an example communication EMS procedure for the Charleston CPW that illustrates the need for both internal and external communications in an EMS.

EMS DOCUMENTATION, DOCUMENT MANAGEMENT, CONTROL, AND RECORDKEEPING

Outline of Element, ISO Requirements, and Purpose and Relationship to Other Elements

Documentation is closely related to document management, and for this reason both elements are addressed in this section.

The ISO 14001 standard contains several requirements for the organization's documentation system and for development of document control and recordkeeping procedures. The basic documentation requirement is for the organization to establish and maintain information that describes the core elements of the EMS and how they interact and to provide direction to the organization. Many organizations will find that this piece of the EMS is one that they already have in place prior to EMS development. As long as an existing documentation system functions as intended, it should work for an EMS also.

The key component for managing documentation is the EMS manual, which describes the EMS elements and how the organization will work to fulfill their requirements. The manual can be viewed as the ISO standard translated into plain language for the benefit of the organization. The manual should contain high-level procedures for each element. These procedures should be very stable, and not likely to be frequently revised or changed on a regular basis. The manual can refer to other lower-level procedures and other documents, such as detailed SOPs, that are more likely to be revised.

This element must also describe procedures on how documents are developed and approved.

How to Get Started

- Review current organizational practices for document control and recordkeeping; an adequate system may already be in place.
- Decide on whether to develop a computerized system for document control and recordkeeping (recommended) or rely on hardcopy filing.
- Assign responsibilities for maintaining the document management and recordkeeping system.
- Develop the document control system to ensure that:
  - documents are regularly reviewed and removed from system if obsolete and are available to employees and are easily located.
  - records are easily located and retrievable, kept for appropriate time periods, and that the system includes all required records (training, monitoring, calibration, reports, management review and other meeting documentation, etc.).

Examples, Tools, and Templates

For a utility that has a computer network, an easy way to get a document control system in place is to use spreadsheet software such as Excel. On a spreadsheet that is available to all appropriate staff, controlled documents can be listed in rows, with columns used to indicate original date, revision date(s), next review date, author, etc. Columns can also be used to indicate if the document is an EMS document, or health and safety, etc. The document names are used as hyperlinks to the actual documents. This simple system fulfills the requirements for a document control system: the documents are readily available in current versions, and obsolete documents are removed from access. A sort on documents checked in the EMS column provides a quick inventory of the EMS documents.

Appendix A.9 contains an example documentation EMS procedure for San Antonio, TX. that illustrates the need to not only clearly define documentation procedures but also to indicate the parties responsible for documentation.

OPERATIONAL CONTROL

Outline of Element, ISO Requirements, Purpose and Relationship to Other Elements

The purpose of this element is to ensure that there are control mechanisms for all significant aspects so that they are properly managed. This element works within the system to make sure that the policy is consistent with the objectives and targets. The ISO requirements for operational controls states that the organization must:

- Identify the operations and activities that are associated with the significant environmental aspects
- Plan these operations and activities to ensure that they are carried out under specified conditions by:
  - Establishing and maintaining documented procedures related to significant impacts

- Establishing and maintaining documented procedures to cover situations where their absence could lead to deviations from the policy and the objectives and targets
- Stipulating operating criteria in the procedures

This element is closely linked to the training element as it will be critical that proper training be provided to the staff responsible for implementing the operational controls. This element is also linked with the preventive and corrective action element as often times corrective actions include updating operational controls.

Considerations for Drinking Water Utilities

The additional focus on public health as an aspect of environmental management for a drinking water utility leads to a focus on product quality greater than there is for other industry environmental management systems. This means that there may be more operational controls that fall within a drinking water utility EMS than in other industry sectors. As most drinking water utilities will already have these operational controls in place, probably as standard operational procedures, they should easily fit into the EMS.

Drinking water utilities will also have greater stress on the critical importance of following operational controls because of the immediate impact on public health if they are not followed correctly. The impact of not correctly following an operational control will need to be communicated well during training and can also be included with the documented operational control to serve as an additional reminder to personnel.

Some drinking water utilities have also run into barriers in implementing an EMS due to bargaining unit agreements or civil service requirements. Most often, such problems arise because the EMS and its related operational controls are seen as causing a change in job responsibilities and job descriptions which may require specific actions within the union or civil service structure. The best advice from utilities that have experienced such situations is to involve union leadership early in the EMS development and implementation process. This will allow them to understand what is happening, why changes are taking place and to make suggestions on how to handle this within their boundaries.

Another issue that some utilities face when implementing an EMS within a government owned utility is that they can not use financial rewards to encourage buy-in for new or changing responsibilities. Utilities have had to find creative ways to encourage buy-in and changes in behaviors such as employee lunches or small giveaways that are not considered compensation.

How to Get Started

The development of operational controls generally follows the following sequence:

- Develop a list of operational controls currently in use.
- Identify the personnel responsible for overseeing operational controls (probably called SOPs).
- Identify, develop and approve any missing operational controls that are needed to appropriately manage significant aspects, and assign responsibilities.

- Document how the above step was accomplished so that additional operational controls can be identified and added in the future as needed.

Implementation Recommendations, Tips, and Suggestions

Be prescriptive with operational controls only as needed. Allow staff to make decisions where possible. For example, if the order of activities during operational rounds is not critical, allow the staff to decide what order the activities are conducted as long as they are all covered within the timeframe required.

There are a couple of ways to approach developing operational controls. Think about utility staff and which of these options would work best:

- Assign different shifts to draft operational control procedures; each shift can then review and comment on another's operational control procedures.
- Because some staff may not be comfortable with writing, a supervisor or someone from the EMS team should meet with operations staff and have them explain the operational control procedures; this person can then write the procedures and have the operations staff review and comment.

After either of these operational control documentation efforts, the procedures should be tested by the relevant personnel to ensure that what is documented is what is actually occurring.

Another recommendation from water utilities that have been through the EMS development process is to look at operational controls not only as an opportunity to make sure that the utility is documenting how things are done, but also as an opportunity for improvement. During operational control development a utility has an opportunity to solicit their staff for their ideas on how operations could be improved. Often employees won't speak up when they have ideas, so it will be important to let them know that the utility is actively seeking their input. It will also be important to follow-up with those staff members who have given input to let them know that their efforts were appreciated and to explain why their suggestion was or was not implemented.

Examples, Tools, and Templates

Appendix A.10 contains an example operational controls EMS procedure and an example operation control tracking table for San Antonio, TX. This example illustrates both the EMS procedure as well as a template that can be used to track operational controls. The template shows the connection between operational controls and monitoring and measurement. This appendix also contains an example of an SOP list and a few sample SOPs from the Elizabethtown Water Company.

EMERGENCY PREPAREDNESS AND RESPONSE

Outline of Element, ISO Requirements, and Purpose and Relationship to Other Elements

Much of the planning involved in developing objectives and targets for an EMS relates to performance and efficiency issues. An important benefit of an EMS also includes requirements to ensure that planning and preparations are made for accidents, natural disasters, intentional attempts to disrupt service and other emergency situations. Emergency preparedness tasks within an EMS support the legal and regulatory requirements, corrective and preventative actions, training and awareness, and identification and communication of significant environmental impacts.

In planning procedures and activities for emergency preparedness, the organization must:

- Address emergencies as inevitable and base planning on this assumption.
- Consider different types of emergencies, including
  - natural disasters (severe weather, earthquake)
  - system failures (power, pumps, controls, wells, piping)
  - intentional malevolent events (consider performing a vulnerability assessment to identify threats)
- Include external partners (police, fire, medical services, media) in planning and exercises.
- Test preparedness and response procedures through tabletop and functional exercises, review and document results, and prepare revisions to ensure continual improvement.
- Understand how procedures relate and link to other elements.

Considerations for Drinking Water Utilities

Water utilities provide a fundamental and essential service to their customers, without which communities cannot function for even short periods. Clean water is essential to public health and economic vitality and must be available to customers at all times with virtually no disruption to avoid serious consequences. Smaller water systems, in geographical scope and volumes of water, can face unique challenges to ensure physical integrity, high purity, nearly continuous supply and service, and limited options to dispose of product not meeting the highest standards. Emergency situations allow little time for actions not previously anticipated and practiced; many water supply emergency scenarios are low probability but high consequence events.

In addition to disruptions in water supply, emergency situations can be associated with the need to release large volumes of contaminated or treated water from the system. Similarly, all parts of a water supply system are critical to human and environmental health, from wellhead or watershed boundary through treatment and distribution. Since 2001, security issues have created new challenges and importance in identifying and testing potential system weaknesses and emergency response scenarios, including the discharge water containing chlorine and other chemicals that can cause environmental problems.

How to Get Started

Since all drinking water utilities in the United States serving more than 3,300 persons must have an emergency response plan in place, this is an excellent starting point from which to review and improve the plan through the EMS process. Assess and document existing emergency response planning through the following:

- Identify internal and external partners, form team(s)
- Identify legal and other requirements and other commitments
- Identify potential threat scenarios (accidents or incidents that could occur)
- Identify plans and procedures to respond to specific emergency conditions
- Document gaps between current and needed response actions
- Develop and revise procedures to address gaps, building when possible on procedures already in place
- Plan, test, and document emergency preparedness and response
- Debrief and evaluate results with team members
- Identify actions to correct system, assign tasks to team members
- Test system and revise accordingly on regular schedule
- Document procedure used for this element
- Repeat at regular intervals
- Include preventative strategies in corrective measures

Implementation Recommendations, Tips, and Suggestions

Developing and implementing an EMS often requires the EMS team, and the entire organization, to change its approaches to defining and solving problems. With regard to emergency preparedness and response, drinking water utilities have already refocused their approaches to meet the Public Health Security and Bioterrorism Preparedness and Response Act of 2002, and mitigate their risks faced from failures due to system vulnerabilities. Although the requirement of the Act is specific to terrorist or other intentional attacks, most utilities have followed the Act's intent that emergency preparedness and response planning not be limited to just intentional acts, but include natural disasters as well. Therefore, using the utility's existing emergency response plan can provide a good platform from which to assess the need for expansion or revision to include accidents, unintentional acts, and infrastructure failure. Some considerations to keep in mind include the following:

- Include organizations identified in emergency plans as early as possible in the planning efforts.
- Use evaluation results of a recent emergency response drill to build on in developing action items for improvement.
- Test emergency response procedures and involve emergency response agencies in drills. Invite observers from other municipalities to ensure readiness drills are rigorous. Seek opportunities to observe emergency preparedness exercises in other municipalities to obtain/exchange ideas.
- Homeland security issues may provide opportunity for funding to conduct comprehensive assessments and improvements.
- Emergency preparedness planning provides an excellent opportunity to identify ways to prevent the causes of potential emergencies rather than planning to respond to the consequences.
- Training of staff is an essential part of emergency preparedness, and testing the effectiveness of the training provides feedback on the effectiveness of the training.

- Anticipation, contingency planning and readiness are critical to responsible management.
- While accidents and emergencies may be inevitable, the frequency and magnitude of consequences can be greatly reduced by good planning and training.
- Training, readiness testing, and continual improvement build confidence in staff and emergency service providers to quickly address what otherwise may be new or complex problem.
- Emergency preparedness procedures, training and practice may help identify needed improvements in other parts of the system.

Examples, Tools, and Templates

Appendix A.11 contains an example emergency preparedness and response EMS procedure for San Antonio, TX. This is a useful example because it illustrates how a utility can incorporate its existing emergency preparedness and response plans into this EMS element and that it is not necessary to develop an entirely new plan for the EMS.

MONITORING AND MEASUREMENT

Outline of Element, ISO Requirements, Purpose and Relationship to Other Elements

While collecting and reporting data are fundamental parts of a water utility's compliance program, ISO 14001 requires that an organization must also have procedures to track "characteristics" of its operations that can have a significant impact on the environment, and to measure progress toward objectives and targets. Additionally, many utilities have measures of efficiency and performance in addition to regulatory requirements. Incorporation of monitoring and measurement programs into an EMS helps ensure that:

- The right data are collected at the right time in the right way to meet requirements
- Data collection and reporting procedures are documented and checked for consistency
- Data used for regular reports to management are collected
- Data needed to support objectives and targets are collected
- Data needed for reporting progress and continual improvement are collected and compiled properly
- The link between monitoring and data needs is clear

ISO has the 14031 standard for Environmental Performance Evaluation that may be, but does not need to be, used to meet 14001 requirements. Revisions in ISO 14001:2002 clarify the importance of calibration and maintenance of equipment on key processes.

Considerations for Drinking Water Utilities

Testing, monitoring and reporting required by regulation is mandatory but may not be responsive to rapidly changing concerns of the public nor to regional or local issues. Additionally, the presence of a single water system for a given area means that residents are typically not aware of how their services and rates compare with other systems, or why. Often, the need for comparative data arises quickly in the event or perception of problems.

However, the similarities shared by water utilities also provide an opportunity to efficiently identify, share and use data collection and reporting systems, and to benchmark performance relative to similar programs across the country, as evidenced by the recent publication of *Selection and Definition of Performance Indicators for Water and Wastewater Utilities* in 2004 by AwwaRF and Awwa, and Awwa's QualServe program. This guidance identifies five categories of measures: organizational development, customer relations, business planning and management, water operations, and wastewater operations.

How to Get Started

Regulatory compliance is at the heart of any measurement program, and can offer a foundation on which to build other data collection efforts. A listing of existing monitoring and data collection activities, and how the data are used, already exists for compliance tracking purposes at virtually all agencies. Using this listing as a point of departure can help ensure that data collection efforts are efficient and are clearly linked to the EMS.

ISO 14001 is intended to manage environmental issues and requires the EMS to address monitoring and measurement for compliance plus significant environmental aspects. Measures that address other issues can be managed within or outside the EMS. If a system of indicators is not already in place, starting simple may be the best approach; early success can help build support for further development.

Terminology

Terminology related to monitoring and measurement is sometimes used differently by different organizations and can be confusing. In general, *issue* or *category* describes the topic or activity that is being tracked. *Data* are the values collected from equipment readings, survey results and documentation of events. *Metrics, indicators,* and *performance measures* are ways of reporting data so that they can be easily understood. For example, customer service is an issue or category; customer complaints are data; and customer complaints per year or customer satisfaction ratings are metrics. These terms are sometimes used interchangeably but in fact have different meanings; make sure there is a clear understanding and definition of each. Some examples of these terms are presented in Table 5.1.

Table 5.1
Monitoring/Measurement Terminology

Term	Example
Category/issue	Regulatory compliance rate
Data	Days above a primary contaminant MCL
Metric/indicator/measure	Compliance rate (percent of days in full compliance with all U.S. National Primary Drinking Water Regulations)
Objective	Consistently achieve 100 percent compliance rate by 2008
Target	Improve compliance rating to 98 percent by 2006, 100 percent by 2007

Additionally, distinctions are sometimes drawn between different types of indicators, notably "output" (or "process") and "outcome" measures. This distinction helps to separate measures of levels of activity and resource levels (outputs) such as miles of pipe inspected, and measures of end results (outcomes) such as water system losses. Outcome and output measures are also referred to as "direct" and "indirect" measures. Make sure the EMS team has a clear and consistent understanding (and use) of terminology. Measurement systems are intended to *improve* communication.

What to Choose

A surprisingly large number of general and specific (to water utility) indicator initiatives have been developed in recent years. The circumstances of each utility will determine where to strike the balance between the simplicity of adopting a packaged set of measures and flexibility to tailor measures to meet specific needs. Selected references are listed at the end of this section.

Understanding the purpose and the audience for the indicators will help guide decisions. If the audience is the public, fewer and more general measures typically work best, while a larger number of more specific indictors is more appropriate for management, contractors and vendors. Usually between five and twenty key measures are manageable.

Remember that a single "higher level" measure can be supported by multiple "lower level" submeasures. For example, "compliance rate" may reflect the percentage of time an agency is in compliance with primary drinking water standards, while supporting measures for specific contaminants can provide additional detail for the internal audience.

Many agencies discover that they need to collect data for several months or longer to create a baseline from which to assess current conditions and set priorities. Establishing a system for collection of data and development of a meaningful baseline are important tasks and may be priority objectives in the early stages of an EMS.

The development of an effective monitoring program may require an iterative process to:

- Define monitoring and reporting needs
  - Define issues to be addressed through reporting
  - Define intended audiences and purposes for reporting
- Define or select desire reporting system and metrics to be used
  - Select from available resources, measurement systems, reporting systems
  - Develop measures to meet unique needs
- Define data availability or ability to collect
- Revise metrics if data not readily or efficiently available
- Set targets based on metrics, if appropriate
- Document procedures for monitoring and measuring.

Implementation Recommendations, Tips, and Suggestions

- Select EMS performance measures that align with and support the utility's organizational goals and objectives.
- Identify, develop, and use metrics that will help the utility measure the benefits of the EMS.

- Keep measures developed for the EMS focused on environmental aspects—this is what is required by ISO 14001—and consider advantages and disadvantages of including other measures that may be used within the EMS. Remember that all procedures identified within the EMS can be included in audits.
- Staff can play a valuable role in identifying, selecting, and testing good metrics. Develop a process that uses their knowledge and expertise so they will "own" data quality and measures.
- Metrics can be used to drive performance—for example, if a goal of reducing customer complaints by 50 percent within six months is set, and employee ratings or incentives are tied to this, results may be seen more quickly. However, make sure the indicator has been carefully thought through and that a new problem is not inadvertently gained while solving an old one.
- For planning purposes, list measurements that the utility *must do* to stay in compliance with regulations. Then list measurements that the utility *should do* (and may do already) to ensure good management. Lastly, list measurements that the utility *wants to do* to tell its story. This approach can help select a useful set of measures.
- ISO 14001 specifically requires that calibration records for monitoring equipment are retained according to procedures; make sure this requirement is addressed and that maintenance records for the equipment used for calibration are also included.

There are many resources available to help organizations select or create indicators. In fact, the large number of resources, each with its own perspective and emphasis, can be confusing. Initial references should include *Selection and Definition of Performance Indicators for Water and Wastewater Utilities (Crotty 2004)*; unless there are reasons to pursue an alternate framework this list of measures should provide a convenient starting list from which to work.

Examples, Tools, and Templates

See reference sources at the end of this chapter.

NONCONFORMANCE AND CORRECTIVE AND PREVENTIVE ACTION

Outline of Element, ISO Requirements, and Purpose and Relationship to Other Elements

This element is an integral part of the act phase (of plan-do-check-act) of the EMS implementation. Preventive and corrective action results in the elimination of causes of nonconformances. The results of preventive and corrective action feed back into the planning phase of the EMS so that procedures are updated as necessary to prevent nonconformances.

The ISO requirements for this element are the following:

- The organization must establish and maintain procedures for
  - handling and investigating nonconformance,
  - taking action to mitigate impacts caused,
  - initiating and completing corrective and preventive action.

- Action taken to eliminate the causes of actual and potential nonconformances must be appropriate to the magnitude of problems and commensurate with the environmental impact.
- The organization must implement and record any changes in the documented procedures.

How to Get Started

The development of a preventive and corrective action procedure generally follows the following sequence:

- Identify the personnel responsible for developing the preventive and corrective action procedure.
- Review reporting procedures and forms that are already in place that can be used or slightly modified to act as preventive and corrective action procedures and forms.
- Review examples. There are some included in the appendix.
- Develop a procedure based on the reviews above and what works best for the utility's operation.
- Have personnel review the procedure and comment on whether or not it will work for them in their day-to-day operations.

Implementation Recommendations, Tips, and Suggestions

One concern that utility staff may have is that there will be repercussions for submitting a preventive or corrective action need. Other drinking water utilities have handled this concern by allowing anonymous submittals. Others have made an effort to acknowledge staff submitting preventive or corrective action needs by identifying the personnel responsible through positive articles in employee newsletters. A utility will need to gauge its staff and what they will best respond to in this situation.

The following are typical steps to follow when a nonconformance has been identified. This may help a utility develop its preventive and corrective action procedure.

- Notification and response
- Root cause analysis
- New actions or procedures
- Verification of effectiveness of new actions or procedures
- Input of changes into management review
- Documentation for audit
- Conduct necessary training

Examples, Tools, and Templates

Appendix A.12 contains an example Corrective and Preventive Action Request/Report Form for the Elizabethtown Water Company. This example illustrates all of the necessary items for corrective and preventive action reporting and tracking such as a description of the problem, identification of the root cause of the problem, the corrective action, verification that the corrective action was implemented, and closure of the corrective action process.

INTERNAL EMS AUDITING

Outline of Element, ISO Requirements, Purpose, and Relationship to Other Elements

Auditing is one of the key components in an EMS responsible for ensuring that an EMS continues to operate as planned, and that it supports continual improvement goals. Internal auditing, whether conducted as the primary audit of an EMS or in support of a third party audit, is a cost-effective method to maintaining EMS performance. Periodic audits of the EMS are required after certification (if obtained) to ensure that procedures and practices continue to work as intended. Internal audits help management ensure that commitments to an EMS are maintained, that objectives are appropriate, targets are being met, and that necessary changes are made over time. Internal audits can be conducted by trained employees of the organization, or contracted auditors.

The internal audit also serves an important function by ensuring that information necessary to conduct a management evaluation is collected. ISO requires management to consider, based on audit results, changing circumstances, its commitment to continual improvement, and the possible need for changes to its policy, objectives, and EMS elements.

The frequency and scope of an internal audit can be defined by the organization under ISO 14001, but generally occur at least once per year between recertification (full third party) audits (or are used in lieu of third party audits if the organization self-certifies to ISO 14001). Many organizations conduct partial internal audits at quarterly or semiannual intervals; though these more frequent audits may have smaller scopes the cumulative result is an audit of the full EMS at least once per year.

The review process used in the internal audit procedure must be documented. Findings from the audits must be documented. Third party auditors will often check to make sure that findings from internal audit reviews by management have been implemented and documented.

Key issues to consider include assuring adequate training for individuals conducting internal audits, ensuring that those conducting internal audits are objective and impartial and documenting the actions and outcomes conducted under this element.

Considerations for Drinking Water Utilities

There is an opportunity for utilities to save money, ensure impartiality and gain new perspectives by trading internal audit services with other utilities in the region.

Ensuring objectivity in internal audits is essential, as has been emphasized by the 2004 updates to ISO 14001. Often those individuals responsible for developing an EMS, while competent to conduct an internal audit and knowledgeable in the details of the EMS, are unlikely to be considered impartial by a third party auditor.

Another opportunity for sharing resources among utilities includes sharing the development of internal auditing procedures and checklists. This approach can further improve the efficiency and ease with which internal auditing services are conducted.

The size of water utilities varies widely, and the scope of audits may also vary significantly. The number of individuals within an organization trained and experienced in EMS auditing will vary as well.

How to Get Started

Planning and preparation for internal auditing should be conducted as part of other EMS development actions, and not left as a final task. Incorporating plans for auditing will help ensure that staff training, documents and procedures are integrated from the beginning. Individuals who will have responsibility for auditing the EMS should be identified early.

Audits are assessments of the completeness and function of an EMS, and help ensure that an EMS, once implemented, remains a priority. Some organizations may have limited experience with system audits, and having an auditor identify areas needing corrective action may not be well received. It is therefore essential that the individuals responsible for internal audits not only be knowledgeable in EMS topics but also have good interpersonal skills.

Staff training helps prepare potential auditees for the processes, questions and outcomes of an audit. Proper preparation can help the organization prepare for and respond to an audit as a team. As noted above, incorporating plans for audit functions into all aspects of EMS planning and development can help ensure that the audits are efficient and effective.

Implementation Recommendations, Tips, and Suggestions

Internal auditors must be impartial and objective, but can work within the organization. Depending on the size of the utility, auditors may audit facilities other than those to which they are normally assigned, or may be from another department in the utility. Auditors may also work for a neighboring facility or be contracted to perform the service.

Training will increase the effectiveness of internal audits by ensuring that all participants understand the procedures and outcomes.

Audit procedures, checklists and reports should be well documented, both to satisfy ISO requirements and to ensure consistency in processes and outcomes.

Defining the scope and frequency of internal audits, relative to ISO requirements and organizational goals, should be discussed with management to ensure the audits meet their intended purpose and that adequate budgets are prepared.

Document management under an EMS may be more rigorous than previous practices. The locations and completeness of documents should be reviewed before an audit begins.

Contractors who are part of operations should be included in EMS training and should be knowledgeable about the EMS, and their specific roles in the EMS.

Audit reports are important checks on how well the EMS works, but may be perceived as criticism of a particular department, program, shift or individual. Awareness of how audit results may be perceived, both by staff and managers, can help ensure that audits strengthen the EMS and the organization.

Inclusion of management at post-audit meetings can help achieve understanding and result in prompt corrective actions.

The recently updated EMS requirements (ISO 14001:2004) clarify record retention requirements associated with internal audits, and also stipulate that internal auditors ensure objectivity in the audit process. These revised requirements should be reviewed and addressed in EMS programs.

Examples, Tools, and Templates

Resources can be found in EPA's EMS handbook for wastewater utilities (EPA 2004).

MANAGEMENT REVIEW

Outline of Element, ISO Requirements, Purpose, and Relationship to Other Elements

Regular top management review is critical to the success of an EMS in any organization. While often viewed as the last stage ("acting" in the plan-do-check-act cycle of continuous improvement), it can also be seen as the first step in setting priorities, objectives and budgets. It is, in essence, the heart of the continual improvement cycle. Without management commitment and investments of time and resources, an EMS project can not start. Review of an EMS program provides a vital opportunity to demonstrate the benefits of the EMS and helps ensure that the EMS is provided continued support.

ISO 14001 requires the management of an organization to review its EMS to ensure it is adequate and effective. Reviews are to be conducted on a regular schedule set by the organization, but generally no less than annually. The review must be documented and must address the following:

- Possible need for changes to policy, objectives, and other elements
- EMS audit results
- Changing circumstances
- Commitments to continual improvement

Management review, though described as a separate element by ISO, is closely linked to internal auditing. As with the audit function, management review also includes an assessment of performance metrics based on monitoring and reporting conducted under the EMS.

Revisions to the ISO 14001 standard (ISO 14001:2004) are more prescriptive in defining issues to be considered in the review process, including:

- Internal audit results
- Legal compliance findings and changes to legal requirements
- Progress toward objectives and targets, including environmental performance
- Continuing actions to ensure continual improvement
- Communication and complaints from interested parties.

Considerations for Drinking Water Utilities

Utility management must balance pressures to keep costs low while ensuring that:

- Product quality and service are high
- New requirements for testing and security are met

- Future demands on system capacity can be met

An EMS can provide new tools to help agency management address, document and report its performance. Since a water utility is fundamentally an environmental service provider, the EMS can be used to demonstrate effective management and justify its expenditures and rate structures.

How to Get Started

An EMS requires that the guidelines for management involvement be developed and implemented. This involvement should include early planning for how management review activities will be conducted. Key decisions will include the following:

- Who will prepare the report, or materials to be reviewed by management
- Who will review reports to top management and who has final authority for content
- Whether the report is planned for distribution beyond top management (to elected officials or the public)
- Who will conduct the management review
- What will be the format of the report, and how detailed will it be
- What issues should be addressed in addition to those specifically required by ISO
- What procedure will be used to follow up and document management decisions

EMS procedures, data collection, and reporting should be considered during EMS planning to ensure that efforts related to this element are efficient and will meet the needs of management.

Implementation Recommendations, Tips, and Suggestions

The needs and expectations of senior management will vary significantly by agency. Early discussions with management about ISO requirements, review cycles, levels of detail, distribution of findings and other factors will help ensure the EMS is valued and effective. Other steps to improve the effectiveness of the management review cycle include the following:

- Provide an objective assessment of system performance; include those things done well and those needing improvement.
- Communicate complex information quickly and effectively through graphics.
- Provide information at a scale that is appropriate for the review; tiering information is a good way to provide high-level summary data while still being able to access supporting details if needed.
- Demonstrate integration of EMS elements through electronic presentations.
- Provide management with objectives for the reviews to ensure that topics of primary interest and importance for management are covered.
- Offer potential solutions to items identified in audits as requiring corrective action.
- Document decisions and action items during the meeting to ensure management support.

- Consider management review cycles more frequently than once per year; regular attention by management will help improve system performance.

Examples, Tools, and Templates

Appendix A.12 contains example documentation of the management review of the Charleston CPW's EMS, which illustrates the importance of documenting the management review, identifying follow-up action items from the review, and how the review is tied to the allocation of resources for the EMS.

INDEPENDENT AND THIRD-PARTY VERIFICATION

Outline of Element, ISO Requirements, Purpose, and Relationship to Other Elements

Once an organization has implemented an EMS conforming to ISO 14001, it can either self-certify that it meets the requirements of the ISO 14001 standard or have an external party conduct the certification. To become certified under ISO 14001 the external party must be an accredited ISO 14001 registrar and the process is referred to as ISO registration or certification. The standard uses the terms "registration" and "certification" interchangeably, but in the United States "registration" is the preferred term.

The American National Standards Institute (ANSI) and the American Society for Quality (ASQ) established the ANSI-ASQ National Accreditation Board (ANAB) for the accreditation of ISO 14001 registration bodies. Minimum registration requirements for ISO 14001 are set by the ANAB, but individual registrars may have additional procedures and policies that an organization must meet. An overview of the registration steps required by the ANAB is discussed in the How to Get Started section below.

Once an organization achieves registration, there is a requirement to perform regular surveillance audits to ensure that an organization's EMS stays in conformance with ISO 14001 and that it is continually improving. The ANAB provides two options for meeting the surveillance audit requirement:

- A 6-month, 1-day surveillance audit cycle with no re-audit after 3 years
- An annual 2-day surveillance with a re-audit after 3 years

Considerations for Drinking Water Utilities

While water utilities do not typically compete with each other for service territory, customers or market share, they face a critical task in maintaining high levels of confidence in the quality of their product and efficacy of their services. Utilities must therefore demonstrate their performance directly to customers or turn to an objective third party for recognition of high standards. Government regulators provide part of this role by overseeing that water quality and testing procedures meets minimum standards for health. However, regulatory agencies do not address issues of efficiency, service, and unique local factors.

Third-party verification of an EMS, with a broader scope than covered by regulations, can help strengthen the credibility of an agency's efforts to achieve high levels of environmental performance. Utilities, typically government agencies but always under public scrutiny, must demonstrate they are well operated and maintained. Third party verification provides credibility to an agency's statements of efficiency and performance. The regular audit cycles conducted to retain certification also provide a strong incentive for a utility to maintain improvements and continually seek improvements, and ensure that these efforts receive priority in budgeting and scheduling.

Justifying the cost of regular third party audit can be an issue for public utilities, and documenting the benefits of an EMS, and certification to ISO 14001, are therefore important. Many firms in the private sector have chosen to implement an EMS but defer certification (seeking so-called audit-ready status), with the reasoning that they achieve the benefits of the EMS without the cost of auditing. This strategy may be appropriate for an organization if

- Their customers do not require certification
- They are not seeking to have their suppliers also operate with environmental management systems
- Their public visibility is low
- Their environmental impacts, both regulated and unregulated, are low
- The potential consequences of environmental impacts are low

Otherwise, third party certification may provide greater return on investment for water supply utilities than some other utilities and many organizations in the private sector.

How to Get Started

ANAB identifies two required stages for ISO 14001 registration (ISO/EIC Guide 66). Some registrars may choose to perform all of Stage 1 onsite while others may divide it into a document or desk audit and a pre-assessment or readiness review. A registrar will perform the audit in the following stages:

Stage 1—Preregistration Audit

The first stage has several goals: (1) planning for the registration audit, (2) gaining an understanding of the EMS in the context of the significant environmental aspects of the organization, and (3) gauging an organization's preparedness for the registration audit.

The Stage 1 audit should be performed at the organization's site, unless an alternative approach can be justified. Adaptation of the registration process to the needs of a very small organization may be justified in particular circumstances.

Stage 2—Conformity Assessment

Stage 2 is performed onsite. The objective is to evaluate and organization's implementation and conformance of an organization's EMS to the ISO 14001 standard.

An organization will be audited to the goals, objectives, and procedures that have been set for their organization even if they go beyond what ISO 14001 requires. An organization must also follow the rules of the registrar.

Implementation Recommendations, Tips, and Suggestions

Most registrars recommend that an organization go through one full year of EMS implementation and one cycle of internal conformance audits and management review before trying to achieve registration.

Accredited registrars are required to have documented procedures and policies which address specific areas of interaction with their clients (e.g., dispute resolution, confidentiality of audit results). When choosing a registrar, an organization should ask to review these procedures and polices, as well as a registrars audit process and make sure that they meet the needs of the organization.

RESOURCES FOR MORE INFORMATION

Accredited EMS ***Registrars***. ANSI-ASQ National Accreditation Board maintains a list of accredited EMS registrars, which is available at http://www.anab.org/.

EMS. See Multi-State Working Group. *The External Value Environmental Management System Voluntary Guidance (DRAFT).* August 29, 2003. Prepared by the MSWG EVEMS Subcommittee. (February 3, 2004; http://www.mswg.org/EVEMSdraft8.29.03.pdf) Stapleton, Philip J., Margaret A. Glover, and S. Petie Davis. *Environmental Management Systems: An Implementation Guide for Small and Medium-Sized Organizations (Second Edition)*. Ann Arbor, MI: NSF International, 2001.

EMS and Local Government. See *Environmental Management Systems: A Guidebook for Improving Energy and Environmental Performance in Local Government*, by Five Winds International/PA DEP, 2004. See also *An Environmental Management System Troubleshooters' Guide for Local Governments,* by Global Environment & Technology Foundation, 2002.

EPA Performance Track. Example of an Annual Performance Report available at https://yosemite.epa.gov/opei/ptapr1_0.nsf/392fb8cedd796f8a85256b8e007da7bb/6c122a09c9dd b0de85256bf10079366f?OpenDocument.

Global Reporting Initiative. See http://www.globalreporting.org/.

ISO 14001. Lexington Group/USAID Office of Energy, Environment and Technology. *Best Practices Guide: Application of ISO 14000 Environmental Management Systems (EMS) for Municipalities*. Undated.

National Performance Track. See http://www.epa.gov/performancetrack/members/ downloads/ final_ ept.pdf.

Peer Center. See http://www.peercenter.net.

CHAPTER 6
RECOMMENDATIONS FOR FURTHER RESEARCH

During the course of the research and development of this report, a number of issues were raised that warrant further study, but are beyond the scope of this project. These issues are listed in Table 6.1 and should be considered for incorporation into AwwaRF's future research agenda.

Table 6.1
Example Research Projects to Improve Efficacy of Drinking Water Environmental Management Systems

Possible Future Research Projects	Objective
Purpose and place of an EMS in drinking water utilities	Identify and assess the benefits and implications that EMS has had on drinking water utility operations, customer service, costs, etc. What would utilities do differently?
Role of an EMS in a utility already participating in other improvement programs	Assess how an EMS fits in with other programs and processes of a drinking water utility (e.g., QualServe, asset management, quality initiatives, etc). Where are the overlaps and inconsistencies? Where are the synergetic relationships?
Aspects and impacts of residuals from chemical treatment processes	Investigate the impacts of chemical residuals, and how design parameters and operational activities can be altered to reduce environmental and other impacts caused by treatment of the residuals, storage, transportation and ultimate disposal.
Aspects and impacts of residuals from membrane processes	Investigate the impacts of brine solutions generated from membrane processes, and how design parameters and operational activities can be altered to reduce environmental and other impacts caused by the treatment and disposal of the brine reject stream. Build on previous research such as Project #607, "Membrane Concentrate Disposal," 1990, and Project #498, "Current Management of Membrane Plant Concentrate," 1997.
Mitigating the impacts from discharging highly chlorinated water from disinfection of new water mains	Build on past research, such as Project #2513, "Guidance Manual for Disposal of Chlorinated Water," 2001, to identify planning activities and operational procedures, for both chemical and non-chemical means of reducing the impact from flushing mains after high-level disinfection. Include the impact from any chemical dechlorination activities.
Environmental impacts of water main relining	Investigate the aspects and impacts of water main re-lining activities. Include different methods, materials and formulations (e.g., epoxy, cement). Identify measures and means to reduce adverse impacts on the environment and community.

REFERENCES

Crotty, Pat. 2004. *Selection and Definition of Performance Indicators for Water and Wastewater Utilities*. Denver, Colo.: AwwaRF and AWWA.

EPA (U.S. Environmental Protection Agency). 2004. *Achieving Environmental Excellence: An Environmental Management Systems (EMS) Handbook for Wastewater Utilities.* Washington, D.C.

EPA (U.S. Environmental Protection Agency). "Environmental Management Systems/ISO 14001—Frequently Asked Questions" [Online]. Available: <http://www.epa.gov/OW-OWM.html/iso14001/isofaq.htm >. [cited May 1, 2005]

NACWA and AMWA (National Association of Clean Water Agencies and Association of Metropolitan Water Agencies). 1999. *Thinking, Getting, Staying Competitive—A Public Sector Handbook*. Washington, D.C.

ABBREVIATIONS

ANAB	ANSI-ASQ National Accreditation Board
ANSI	American National Standards Institute
APWA	American Public Works Association
ASQ	American Society for Quality
EMAS	Eco-Management and Audit Scheme
EMS	environmental management system
EPA	U.S. Environmental Protection Agency
ISO	International Organization for Standardization
LRC	Local Resource Center
NACWA	National Association of Clean Water Agencies
NBP	National Biosolids Partnership
PAC	project advisory committee
PEER	Public Entity Environmental Management System Resource
SOP	standard operating procedure
WEF	Water Environment Federation

IWA Publishing's authorised EU representative for General Product Safety Regulations is Diane D'Arras, 15 rue Duret, 75116 Paris, France, e-mail: safety@iwap.co.uk.

Printed and bound by CPI Group (UK) Ltd, Croydon, CR0 4YY
17/09/2026
02218810-0009